COMMON SENSE WIND ENERGY

COMMON SENSE WIND ENERGY

California Office of Appropriate Technology

Brick House Publishing Company, Inc.
Andover, Massachusetts

Published by Brick House Publishing Co., Inc.
34 Essex Street
Andover, Massachusetts

Production Credits:
Design: *William McGuire*
Illustrations: *William McGuire*
Editing: *Robert Kahn*
Produced by Mike Fender

Copyright © 1983 by Office of Appropriate Technology, State of California. All rights reserved.
Printed in the United States of America.

Library of Congress Cataloging in Publication Data
Main entry under title:

Common sense wind energy.

  Bibliography: p.
  1. Wind power. I. California Energy Commission.
II. California. Office of Appropriate Technology.
TJ825.C65 1982 621.4'5   82-14765
ISBN 0-931790-38-7 (pbk.)
ISBN 0-931790-41-7

CONTENTS

Introduction	7
Chapter 1: OVERVIEW	9
Using Wind Power	10
Chapter 2: FUNDAMENTALS OF ENERGY NEEDS	13
Pumping Water	14
Guide 1	15
Producing Process Heat	17
Guide 2	18
Running Appliances	19
Conserving Energy	20
Chapter 3: FUNDAMENTALS OF WIND SYSTEMS	23
Guide 3	24
Pumping Water	25
Producing Process Heat	28
Generating Electricity	29
Being Your Own Utility	30
Selecting Equipment	32
Chapter 4: FUNDAMENTALS OF SITING	37
Ensuring Safety	37
Maximizing Energy Production	38
Doing Your Own Wind Study	43
Guide 4	44
Chapter 5: DEALING WITH ECONOMIC, LEGAL, AND SOCIAL ISSUES	49
Economics	49
Legal Issues	52
Social Issues	54
Guide 5	55
Chapter 6: BUYING, INSTALLING, AND OWNING A WIND SYSTEM	61
Typical Installation	61
Typical Maintenance	62
Conclusions	63
Case Studies	64
Appendix 1: Power and Energy Requirements of Appliances and Farm Equipment	67
Appendix 2: Electrical Wind Machine	71
Glossary	75
Bibliography	81

ACKNOWLEDGMENTS

This edition of *Common Sense Wind Energy* is based on a book by the same name written by Jack Park of Helion, Inc. and John Obermeier, P.E. of Otech Engineering, Inc. The original edition of *Common Sense Wind Energy* was published by the Office of Appropriate Technology and California Energy Commission in spring 1981. This edition has been revised and updated since its original appearance.

A great many people helped to produce both the original and Brick House editions of *Common Sense Wind Energy*. Special thanks to Bob Judd, OAT's director, for his support of the project, to William McGuire for designing and illustrating the book, and to Gigi Coe, OAT's Deputy Director, for overseeing its preparation. Robert Kahn edited the original and present versions. Remy Ceci, OAT's wind specialist, and Vicki Butler, her assistant, contributed to the current edition.

The book's manuscript was carefully reviewed by a number of wind specialists including Don Bain, Rick Bennett, Paul Gipe, Elena Jones, and Jim Lerner. Among the OAT staff who helped to produce the book were Brett Parent, Marian Wilcox, and Cheryl Yee. Photographs were taken by Joe Carter and Mush Emmons. Production coordination was contributed by Mike Fender.

Jack Howell
Publisher
Brick House Publishing Company

Introduction
Governor Edmund G. Brown, Jr.

In many parts of the country, wind is a "common sense" source of energy. It is clean, economical, and plentiful. Wind machines can now produce electricity directly for home use or for sale to utility companies. Together with the opportunities before us to fully develop solar, cogeneration, biomass, geothermal, and other alternative energy sources, wind is integral to establishing a secure energy future for the United States.

Common Sense Wind Energy should dispel the myth that wind energy is an exotic idea. This book is written to show how many people can reap the energy and financial benefits of wind. It demonstrates that the simple prerequisite is a parcel of land with a steady wind flow. By measuring the speed and consistency of wind, a potential wind machine owner can accurately predict just how much energy he or she can expect to generate. After this determination, federal and state tax credits and mandated utility buy-backs of surplus power assure the sound business sense of a wind energy investment. It is these buy-back regulations in fact, under the 1978 Public Utilities Regulatory Act (PURPA), that allow wind machine operators to be their own utility companies.

We are at the beginning of a new era of electrical innovations from wind energy. In the past decade, technological innovations have emerged from individuals, industry, and government research and development. There are over fifty wind machine manufacturers in the United States, making machines that generate anywhere from 3 kilowatts, which is enough to power most household appliances, to 4.5 megawatts, which is enough to power a small city.

States have become more aggressive in supporting and promoting the installation of wind machines and, where feasible, wind farms are being built to make an even larger contribution to our energy demand. In California by the year 2000, we project that over 7,700 megawatts of wind-generated electricity will be produced, equalling about 10 percent of the state's electrical consumption. Soon, we will have wind machines generating electricity at state parks and facilities. And, annually, the industry for both individual and wind farm manufacture and installation has doubled in size — against the recessionary tide that has hurt other small businesses.

Whether for economic reasons, commitment to renewable energy sources, or for the exhilaration of having achieved freedom that comes from disciplined understanding and wise use of natural resources, I hope you will read *Common Sense Wind Energy* as an invitation to join in this adventure of wind energy development. The resource is there and we now have the technological capability to use it wisely and profitably.

6/1/82

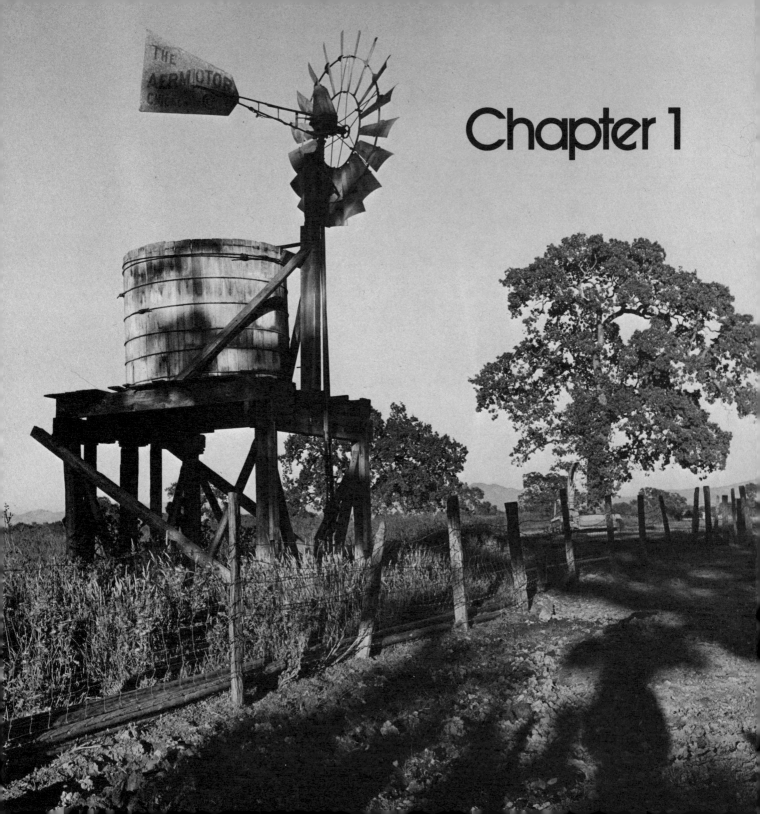

Chapter 1

OVERVIEW

Americans have two options available in dealing with rising energy costs and our concern about energy supplies. We can make more energy available by using what we have more efficiently or produce energy ourselves. Fortunately, we can do both.

Increasing energy efficiency and using less is called conservation. Conserving energy accomplishes two immediate objectives among many others:

- It helps assure that what energy is available can serve more consumers, and
- It helps reduce dependence on conventional, non-renewable energy sources.

Energy conservation is the quickest, cleanest, safest way to "make more energy." It is also the most economical way. But energy conservation will not satisfy all of our energy needs. Because of increasing population and a growing economy, we will need to tap additional energy sources.

New energy sources don't demand the construction of new, large power plants. Instead they can mean increasing individual use of renewable sources such as solar, biomass and wind energy.

This book has been prepared by the Office of Appropriate Technology as a guide to understanding one way of making more energy from a source that is available in some areas — wind energy. Parts of the U.S. are rich in wind energy; others are not. This book will help you understand how to determine if your particular area is suitable for harnessing the wind's energy. It will then help you understand how to choose appropriate equipment for a wind energy system that will fill your needs.

America's energy needs during this century have been satisfied primarily by oil, coal, natural gas, and hydroelectric power. The development of these energy sources, together with nuclear power, has historically held back development of wind energy. One exception was the widespread use of wind in the West and Midwest. There, before World War II, windmills were used to pump water and produce electricity in rural areas.

Today, with electric rates rising dramatically, interest in wind energy is growing. As you seek to determine whether the wind can work for you, keep in mind that whether the wind can provide you with cost-effective power depends on the quality of your local wind resource and your energy requirements.

If you have access to abundant wind you'll find that it can perform several important tasks for you.

All together, consumers' needs tend to group into three areas:

- pumping water,
- process heat, and
- appliance power.

Water pumping is perhaps the energy use warranting the most immediate reliance on wind energy. Energy is needed to move water for many purposes, including lifting well water for home and livestock use and for irrigating crops. Fortunately, many of these applications are in areas rich in wind energy.

As defined in this book, **process heat** is a term used to describe production of heat for virtually any purpose. Heat is needed to warm living spaces and to heat domestic water. Heat is needed in agriculture to prevent frost, dry crops, sterilize equipment in dairy and other operations, and to produce steam for food processing such as canning tomatoes and refining sugar. Heat is needed in chemical, aerospace, and electronics industries. Process heat is even needed at the local laundromat. Many of these process heat users are in windy areas.

Appliance power covers just about everything not mentioned above. Mostly, however, it means running appliances for domestic and small industrial uses. Refrigerators, dairy milk chillers, televisions, and other not-too-heavy users of electricity can be powered by wind energy.

USING WIND POWER

Before proceeding with planning your wind energy system, you must organize the steps you will take. Here is a convenient approach for planning your wind system:
- Determine how much wind energy is at your site.
- Figure out how much of the energy you use can be supplied by wind.
- Select a suitable wind system that matches the wind resource to your energy needs.
- Proceed with planning the installation, financing, tax credit applications, and resolution of social and legal issues.

It may be that you do not live on a windy site. It may be that your company is planning to find sites for commercial development of wind energy. In either case, searching for windy sites — called **wind prospecting** — is necessary before you can plan a wind system.

Once windy sites are located, the amount of energy available must be determined. This is called **site analysis**. Occasionally a site that was thought to be windy during prospecting will be disappointingly low in wind energy. Often, the opposite is true — a site thought to be marginal is far more energetic than expected. Careful site analysis reveals such information and is discussed in detail in a later chapter.

The second step, determining your energy needs, can be a difficult but rewarding task. Often, the potential to conserve becomes apparent while doing an audit of your energy needs. For commercial enterprises, it is usually much simpler to perform an energy audit — much less emotional attachment is affixed to any particular energy use. Areas of conservation potential are more easily identified.

In any case, it is not wise to assume that all of your energy needs will be met by wind energy alone. Wind energy users in the 1940s may seem to have had their electrical needs supplied entirely by wind energy, but in fact many midwestern wind energy machines had a kerosene or gasoline back-up generator to charge their batteries during long windless spells. One alternative to the back-up generator was to load the dead windcharger batteries into the wagon and head for town. The local blacksmith usually had a generator around to boost batteries while the farmers did their shopping.

Those days are probably gone forever. Today, realistic wind system planning pegs wind power supply percentages at between 25% and 75% for individual wind systems.

Selecting a suitable wind system requires either contacting a nearby dealer of wind energy equipment, or contacting a variety of manufacturers of wind machines, towers, batteries, inverters, and a host of other devices. At the end of this book is a list of manufacturers to help you select the equipment best suited to your needs. Several good books are also available that discuss wind energy matters in much greater depth than is possible in this book. They are listed in the Bibliography.

The key to satisfaction with a wind installation starts with your understanding of what any given wind system will and will not be capable of doing. This means understanding what you want the machine to do for you, then taking care to plan an installation that will perform as expected. To succeed, you must understand the principles involved in wind energy, in your energy use, and in wind system planning and installation. The following chapters are organized to help you understand these principles. We shall start with you and your energy needs.

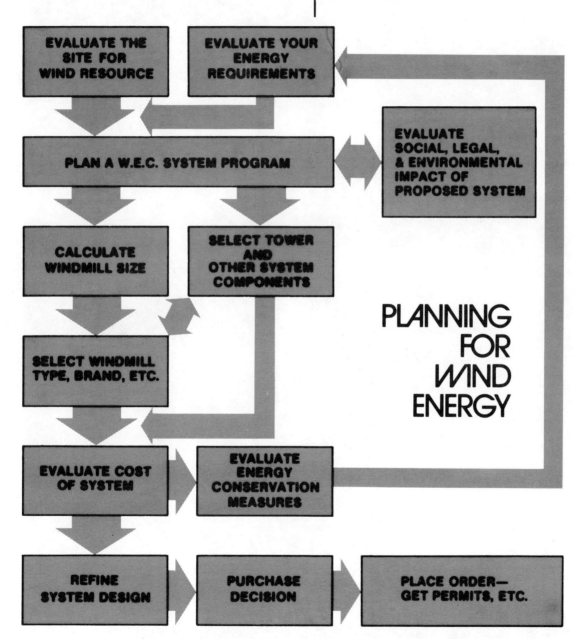

PLANNING FOR WIND ENERGY

Chapter 2

FUNDAMENTALS OF ENERGY NEEDS

Suppose you are holding an ordinary light bulb in your hand. Look at the marking on the bulb and notice that it is rated at 100 watts. That is 100 watts of electrical power: the amount of power the light will consume when you install it and flip the switch on.

Suppose you leave the switch on for one hour. The light will have consumed 100 watts for one hour, or 100 **watt-hours** of energy. Energy can be calculated by multiplying power by time: if you leave the light on for a day, it will have consumed 24 x 100 = 2400 watt-hours, or 2.4 kilowatt-hours. One **kilowatt-hour**, abbreviated **kWh**, is 1,000 watt-hours. Your electrical power meter counts up the kWh you use. The electric utility company reads this total kWh and multiplies it by the cost of electricity to arrive at the electric bill you pay. Therefore, if electricity costs 5¢ per kWh, leaving the 100 watt light on for 24 hours costs 5¢ x 2.4 = 12¢. It costs you 12¢ to leave that light on all day. Leave it on all month at that rate and you spend about $3.60.

Another example of power and dollar calculations: Suppose a one horsepower (hp) pump can lift 300 gallons each hour from your well. One horsepower equals about 750 watts of electrical power. Suppose that you must run that well pump 30 hours each month to meet your needs. Then 750 x 30 = 22,500 watt-hours, or 22.5 kWh. At 5¢ per kWh, that would cost about $1.13. The well will have pumped 30 x 300 = 9,000 gallons of water out of the ground during that month.

With these types of calculations in mind, one of the tasks you must complete is an estimation of your energy needs. This is done to calculate what size wind machine you need and to predict its performance. Start by establishing project goals. Examples of these goals would be:

- To pump half of your irrigation water needs out of the ground with wind power, the rest with utility-supplied electricity.
- To help heat a house or greenhouse during cold, windy, winter months, and reduce the cost of heating household water the rest of the time.
- To offset the cost of chilling milk in a dairy milking parlor.

These and other uses of the wind's energy are typical of the goals you may select. Some goals may be combined so that one wind system can be planned to serve several needs; other goals are best kept separate so that wind systems designed for particular needs are not expected to perform other tasks for which they are not well suited. For example, water may be pumped with electricity produced by the same wind machine that is used to power a household. However, if the pumped water is needed far out in a pasture for stock watering, one of the familiar farm water pumpers might be best while a small, economical wind electric generator serves the household.

After defining your goals, calculate the actual energy you expect to demand from the wind system. This amount might be 50% of your total need, in which case you assess your total need and use half of that value for planning the wind system. The actual calculation process for each of several types of energy uses is explained in Guides in this chapter. The following sections discuss the less technical aspects of each of these.

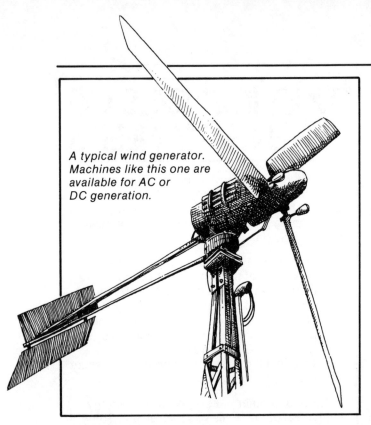

A typical wind generator. Machines like this one are available for AC or DC generation.

PUMPING WATER

Moving water is bound to require some effort and take some time. Lifting 550 pounds of water (about sixty-five gallons) one foot off the ground in one second requires one horsepower. That's the amount of power required if you merely hoist that much water aloft in a big bucket. If you try to pump that much water through a one-foot long pipe, the amount of power required will increase greatly.

It doesn't take much imagination to realize that you probably won't get sixty-five gallons of water through a garden hose in one second, even if the hose is only a foot long — not even 100 garden hoses will do. That's because too many molecules of water are forced through too small a hole too fast. You can see, then, that the power required to move water is determined by how high you wish to lift it, how quickly, and the size of the tube (both in diameter and length) that it will pass through. The information presented in Guide 1 will help you to estimate the water pumping power you need.

Remember that energy equals power multiplied by time. Once you have estimated the power required, then calculate the energy needed. Using the example in Guide 1, you have calculated that lifting water from a certain depth in your well at a rate of 300 gallons per hour (gph) will require about one horsepower, or 750 watts of electrical power. You need 9,000 gallons each month, so your monthly energy demand is calculated by finding the number of hours the well must pump: hours = 9,000/300 = 30 hours. Then electrical energy equals 750 x 30 = 22.5 kWh. If you want 50% of that energy to be supplied by the wind then plan your wind system to provide 11.3 kWh.

Conservation is as important with wind energy as it is with other types of energy. When pumping water, there are two ways to conserve: one is with individual care, the other is by careful wind system design. The first is to reduce the need for water. For stock watering, automatic watering devices are available that reduce algae build-up and evaporation, thereby reducing water demand. Obviously science has not invented a means to prevent farm animals from needing water. Yet.

Conserving water is the best way to reduce the size of the wind system because you reduce the amount of water it must pump. Once you have conserved all you can, proceed to plan the wind system needed to pump the required amount of water. During the planning phase, you will be confronted with decisions about what size water lines to install and how many bends, elbows, joints, and other widgets to install in the plumbing. Each of these decisions affects the final outcome; smaller pipe and more joints, bends, and elbows all add to the power required to pump water. The trade-off is to pay for larger pipe (for reduced power requirement) or for a larger wind system to pump water through the smaller pipe. In most cases, the larger pipe is cheaper in the long run.

ESTIMATING POWER REQUIREMENTS FOR WATER PUMPING

GUIDE 1

A simplified method of estimating your power needs for water pumping requires that you know how high you want to pump the water, and how fast. The diagram illustrates a typical water-pumping application where water is drawn from below ground to a storage tank above ground. The total height is the well depth plus the storage tank height. How fast you pump the water is called the flow rate and is usually measured in gallons per hour. You normally determine your required flow rate from your livestock or irrigation requirements.

You will always need more power to drive the wind machine than the power required to just pump the water. This is because of less than perfect windmill efficiencies. Typical inefficiencies for wind-driven water pumpers are accounted for in the chart.

GUIDE 1

For example, suppose that you want to pump water from 250 feet below the ground into a tank 50 feet above the ground at a flow rate of 300 gallons per hour. Use the enclosed chart to estimate your required horsepower. First, find your required height of 300 feet on the horizontal axis. Then go up to the 300 gallons per hour line and go left to read one horsepower on the vertical scale. For the range of flow rates that can reasonably be expected from this type of wind machine, this graph provides a rough estimate of power requirements.

Information supplied by manufacturers of commercially available conventional water pumpers may be used for comparison of power requirements. Remember that this procedure only establishes general estimates. Conservative selection of a windmill should allow for increasing power for future growth or as a safety factor. Increase of half again the power estimated in the example are not unreasonable.

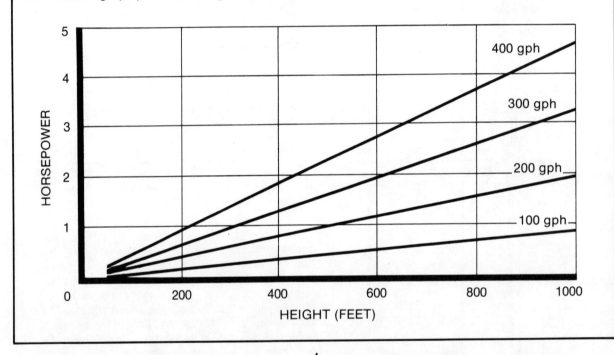

A number of methods are available for estimating power requirements to move water. For example, you can simply measure the time your well pump currently runs. Read the data plate on the pump motor to find its horsepower rating. Factor in expected growth in water demand, especially if readings are taken in winter or if a new baby is on the way. If an electric meter is hooked to a field-installed well pump, simply consult your bill files or your electric utility to arrive at an energy-use summary for that well pump.

Determining the energy required for moving water is only the first step in the decision-making process which may lead you to purchase wind energy equipment for this task. Your next step is to conduct a site analysis (described in Chapter 4). In the interim, you may want to examine some of the other contributions wind energy can make.

PRODUCING PROCESS HEAT

Process heat is usually thought of as heat used in chemical factories, steam plants, laundromats, and such. Here it also is taken to include heating buildings since generating heat by wind power is the same for all heat needs.

Heat is energy in a form easily understood. We do not usually think of an elevated water tank as stored energy, but it is — especially if the tank's water was lifted 200 feet out of the ground to rest ten feet high. We think of heat as energy because we are familiar with burning natural gas or glowing electrical heaters. We are also familiar with the bills we must pay for their pleasures.

The usual reference to heat energy is written in **British thermal units (Btus)**. One Btu is the amount of heat energy it takes to change the temperature of one pound of water by one degree Fahrenheit (°F). Thus, one Btu is taken from a pound of water if its temperature is lowered by one degree F; one Btu is added if the temperature is raised by one degree.

Since windmills are not rated in Btus but rather in watts of kilowatts, you might expect some problem. Fortunately, 3414 Btu equal one kWh's worth of electrical energy. The conversion is simple: divide the number of Btus you need by 3414, and the answer is the number of kWh of wind-generated electricity you need.

Calculating process heat energy is shown in Guide 2 as two separate types of calculations: one is to determine energy required to raise the temperature of some heat-absorbing substance like water, concrete, or rocks, and the other is to estimate heating requirements of buildings — homes, in the example. To obtain highly refined estimates, you should consult a qualified heating engineer or architect. The formulas and graphs given in the Guide are useful for reasonable estimates, but will not result in a truly accurate forecast of your energy needs. But if you are shooting for 50% or so of your energy requirements to be supplied by wind power, then an exact estimate is not really required for sizing the machine. On the other hand, reasonably reliable estimates may be required to satisfy the banker you plan to seduce into financing the project. Care should be taken in any estimate. Contract qualified help if the project warrants it.

Heat might be supplied by a wind machine only when it is windy and cold. If this is the case, your home's heating needs will be increased because cold winds carry heat away faster than still air. Also, any building's cracks and leaks become little windy canyons during cold winter breezes.

Wind machines that provide space heat are called **wind furnaces**. These systems are typically wind electric generators wired for heating purposes. In some cases these generators can supply electricity directly to the building wiring; in other cases the wind-generated electricity is wired to heater probes installed in water tanks, heating ducts, concrete slabs, or rock beds.

There are many opportunities to conserve energy in home heating. Reducing the need for heat reduces the size of wind machine required. The classic ways to reduce heating needs are to lower the thermostat and either insulate or insulate more heavily. Several excellent books have been published describing ways to improve the thermal performance of a building, water heater, or other heat user. Plan to review these books before investing in a wind system.

Heat Storage Tank

GUIDE 2

BUILDING AND PROCESS HEAT ENERGY

Everywhere in nature, thermal energy seeks to move from warm to cold places. For example, a warm house loses heat to the cold outdoors. Conservation-minded occupants insulate the house, thereby slowing down the rate at which the heat is lost to the outside; but whatever its rate, heat loss is inevitable.

To estimate the amount of energy it takes to heat a house, you need to know the rate at which heat is lost. Heat-loss calculations can be completed by anyone armed with a pocket calculator and plenty of patience. A simple procedure based on tabulations of many heat-loss calculations is presented here.

The accompanying graph suggests the amount of energy required to heat a house, from the uninsulated to well insulated. Heat energy requirements are shown for no wind condition, and for a wind of 15 mph. Solar heat gain on typical walls and windows is estimated. Keep in mind that the graph presents a typical range; it excludes the best and the worst in building design.

The energy units are presented in Btus per degree day per square foot. Degree days are a measure of the amount of heat it takes to maintain comfort conditions in a given locale. To calculate the energy requirements for your house, you need to know the degree days at your location, the floor area of your dwelling in square feet and its relative insulation and construction quality.

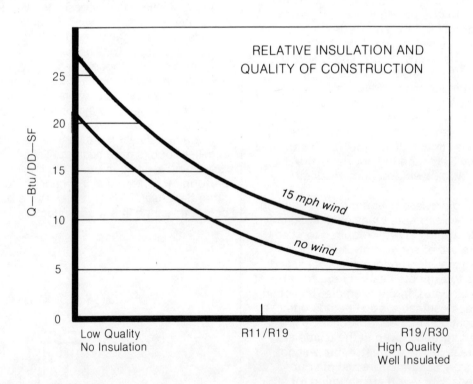

GUIDE 2

Example: *Take an average insulated house of 1500 square feet located in Colorado Springs, Colorado. How much heating energy is needed to maintain this house at 65° F during January?*

Solution: *Assume January to be a windy month (14 mph average) and move to a point midway along the horizontal axis of the graph (to represent medium levels of insulation). Now move up the scale between the no-wind and 15 mph wind condition. Find "Q" on the left-hand scale at 10 Btu per degree day per square foot. (Colorado Springs has 1128 degree days during January.) Now calculate the required heat with "E" representing heating energy.*

$$E = 10 \text{ Btu/DD-SF} \times 1128 \text{ DD} \times 1500 \text{ SF} = 16{,}920{,}000 \text{ Btu}$$

or in terms of electric units, the heat required is:

$$E = 16{,}920{,}000 \text{ Btu}/3414 \text{ Btu/kWh} = 4956 \text{ kWh}$$

This amount of energy could be supplied by a fairly large wind generator.

One method to balance constant residential demand for energy with the unevenness of wind availability is with rock bin thermal storage. Excess heat energy produced during windy spells can be stored in the rock bin for later use and extracted when desired by blowing air through the bin and into the residence.

Example: *How big a rock bin would it take to store three days' worth of heat in January for our Colorado Springs house?*

Solution: *River rocks average about 150 pounds per cubic foot and have a specific heat of about 0.2 Btu per pound per degree F. The heat energy required for three days (one tenth of a month) is 1,692,000 Btu. Rocks can be heated over a temperature range of 80° F to 250° F, or 170° F. The amount of rock required is:*

$$\frac{1{,}692{,}000 \text{ Btu}}{0.2 \text{ Btu/lb.} - °F \times 170°F} = 49{,}765 \text{ pounds}$$

$$\frac{49{,}765 \text{ lb.}}{150 \text{ lb./Ft}^3} = 332 \text{ cubic feet}$$

If this was put in a cubic bin, the container would be seven feet on each side.

RUNNING APPLIANCES

The energy needs of appliances are estimated in much the same way as the energy requirement of a well pump. Appendix 1 shows a table of statistical averages for power and energy requirements for a variety of appliances. This list by no means covers every device you may use, but it does include the most common residential appliances. For commercial and agricultural appliances, consult dealers of the products to find their energy specifications.

A typical strategy for predicting energy consumption is to read previous energy bills. To do this, consult your files for past bills or request a summary of your patterns of energy use from your utility company.

With this summary, you can use the values given in Appendix 1 to add or delete energy-using appliances as your situation demands. Thus, if you plan to add one more television set to your list of energy gadgets, consult the Appendix and adjust your energy summary accordingly. Perhaps you might plan to

A three-blade "eggbeater" Darrieus rotor. Invented in 1926, this design is finally being proven.

eliminate an appliance; then subtract the appropriate estimate based on the table provided from your summary.

In the absence of any available energy data, generate your own by simply listing the appliances you plan to own, and estimate their energy consumption. A typical American household uses an average of 750 kWh per month; even in warmer climates homes tend toward this average — the electricity they don't use for winter heating, they make up for in summer cooling.

CONSERVING ENERGY

Appliances, as energy users do not constitute even half of the nation's total energy consumption. Climate control — heating and cooling — are the major users, although you might think of an air conditioner as an appliance. Like climate-control conservation measures such as improved insulation, lowered thermostats in the winter and raised thermostats in the summer, appliances also offer many conservation opportunities.

Take refrigerators and freezers for example. It has been estimated that if all of these appliances used throughout California were to be improved from a typical monthly average consumption of about 200 kWh to a much lower consumption on the order of fifty kWh (a value quite easy to attain), California's **peak load** (time of greatest electrical demand) would shift from summer to winter.

To understand seasonal and daily peak loads, consider the following. During the evening and overnight hours, American homes are largely "powered-down"; energy use is low during these hours. The same goes for much of the commercial and industrial sector. But about 8 a.m. each morning, everybody is up, running every imaginable appliance. The result is a terrific surge in power demand. Usually, between noon and early evening, the electric company registers a peak in power demand. Just about all available power stations are working at full capacity. If they were not all generating, the toaster you just switched on might dim your kitchen light. If you were a small computer consulting firm, you might find that such a voltage drop would cause your computer to print out garbage.

The peak situation is a central problem faced by electric utilities. The size of the peak power demand is the primary factor in deciding to build new generating facilities. As the peak demand grows, the need for new facilities increases. Appliance users play a key role in determining peak power demand. Through careful choice of appliances (and the time of day they are used), consumers can control peak demand.

California, the Southwest, and states south of the Mason-Dixon line are generally summer peaking states. That means that the largest peak in power demand occurs in summer on the hottest days. Cooling is a major factor, because everything that has a cooling job, including refrigerators, has to work harder when it is hot.

Conservation strategies for appliances involve two areas: first is reducing the peak, and second is reducing the overall power required. Peak demand reduction comes about by not turning on appliances during peak load hours. Overall power reduction comes about by improving the performance of appliances.

You might not think of leaving your refrigeration equipment off as a particularly helpful energy conservation strategy. After all, the ice will all melt and just have to be refrozen later. That is quite true, but not entirely appropriate here. Leaving an appliance off does not necessarily mean leaving it off all day; simply scheduling appliances to be off during, say, ten minutes out of each hour will have no measurable effect on the contents of a refrigerator or air-conditioned home. But it will, if carefully scheduled to coordinate with similar activities in other buildings, have a tremendous impact on the ability of the electric company to serve the needs of a growing clientele.

Consumers can participate in a **load management** program either individually, or in cooperation with their utility. For example, utilities are now offering reduced rates to consumers who allow them to install load management devices. Consumers can purchase cheap clock timers on their own and plug appliances like refrigerators, freezers, and such into these clocks. By preventing them from coming on during peak hours — breakfast, lunch, and dinner hours, for example — a contribution is made toward reducing the peak.

Reducing power demands of appliances means improving their performance. Many new appliances already are greatly improved: refrigerators are now available which draw less than half the usual energy needed to keep food cold. Appliance performance is improved not only through its design, perhaps through better insulation, but also by altering the way you use it. Running only full loads in washing machines, and using the coolest possible water in the wash and rinse cycles are examples of this latter strategy.

Wind power can also make a contribution to load management, especially if the wind energy at your site is available during daytime hours. Thus, while you are assessing your energy needs, you are also thinking in terms of the wind resource. Electric utilities are gearing up to encourage you to use appliances during night hours. Your electric rates will probably soon be cheaper at night. That means that wind energy available during the day will be worth more to you and to the utility.

The next chapter explores the fundamentals of wind energy conversion systems and assessing the wind resource. It's a good idea to think in terms of the entire range of options available to you, from conservation to installation of wind equipment. This means you must be familiar with all aspects of wind energy conversion — everything from your energy needs to the ability of your wind resource to provide for those needs and the selection of wind equipment capable of withstanding the hazards of weather and ownership.

Chapter 3

FUNDAMENTALS OF WIND SYSTEMS

Wind energy is available in the form of speeding molecules of air. Wind machines convert this form of kinetic energy into more useful forms such as electricity by slowing these molecules down.

The secret is knowing how to do it. If you build a fence across the direction of a wind, you will hardly slow the wind down; some of the air will simply pile up in front of the fence and form a ramp for the rest of the wind to slide over. A tall tree, on the other hand, is quite effective at extracting power from the wind. A tree flexes, transferring the wind power into the bending and twisting of its limbs and trunk. Trees, however, are not practical as wind energy converters for consumers. Some clever experiments have been tried to harness swinging trees to pump water; the results have been disappointing at best.

Wind energy conversion systems — windmills, as we call them here — perform their task by slowing the wind down. Like the fence, windmills cannot stop the wind. When operating at their peak efficiency, windmills slow the wind directly in front of them down to about one-third of the original wind speed. Any more or less than this, and the windmill is not operating at its peak efficiency, as a general rule.

Measuring the wind speed in front of and behind a windmill, however, is not a particularly useful way to measure the efficiency of a wind machine. Studying manufacturers' performance charts will give you some clue of the candidate windmill's ability to supply your energy needs. Here you begin the three-way matching task of selecting equipment that is capable of matching your site's energy resource to your energy needs. Because of this site/need/windmill match, you cannot simply go out and purchase whichever windmill suits your fancy or budget. To do so is to increase the possibility of a mismatch. To establish a good match is, however, a reasonably simple and even enjoyable task. The many factors that affect such a match, at least from the windmill's point of view, are discussed here.

Wind power can be calculated in the same terms we use to talk about electric appliances — watts, or even horsepower. It's not technically feasible to measure wind power, but you can measure wind speed and calculate the power. The equation is given in Guide 3, but the idea is that wind's power is proportional to wind speed cubed: mean wind speed multiplied by itself three times (speed x speed x speed). This cubic relationship means that twice the speed produces eight times the power available. Based on this principle, two actions become important in your quest to harness wind energy. First, find a site with the most wind energy available, which generally means highest wind speeds, and second, select a wind machine that's both within your budget and capable of extracting the most power out of the wind.

A number of factors affect the efficiency of a wind system; not all of them are associated with the windmill itself. Some of them are associated with the height of the tower, size of wires that carry electricity from the generator to the load (remember the problem with small water pipes mentioned earlier?), and the efficiency of various devices you may need in between the generator and load. The intended purposes you plan for wind energy will dictate your design of the system. Because there are so many factors, wind systems are discussed here in terms of planned use, starting with water pumping.

GUIDE 3

POWER IN THE WIND

The amount of power in the wind is dominated by two factors:
1. Instantaneous power varies with the cube of the wind speed, and
2. Wind speed variability influences the long-term mean power available.

The power in the wind is one-half the air density (Kg/m^3), times the area (m^2), times the cube of the wind speed (m/s). Expressed as an equation this is:

$$P = \tfrac{1}{2} D A v^3$$

where P is the available power in watts (W);
D is the air density in kilograms per cubic meter (Kg/m^3);
A is the cross-sectional area facing the wind flow in square meters (m^2);
v is the instantaneous wind speed in meters per second (m/s).

It is convenient to talk about available wind power in terms of power per unit area facing the direction of wind flow. In this case, the power equation can be written:

$$P/A = \tfrac{1}{2} D v^3$$

where P/A has units of watts per square meter and is often referred to as the **power density.**

The equation above shows that the power density varies directly with the air density (doubling the air density doubles the power) and varies as the cube of the speed (double the speed power increased eight times). From the figure, it is clear that air density variations are a small factor in wind power compared with the effect of wind speeds.

The effect of wind speed variability on mean power can be demonstrated by comparing two simple examples. Consider a site where the wind always blows at 6 m/s. In this case the long-term mean power density is:

$$\tfrac{1}{2} \times 1.22 \times 6 \times 6 \times 6 = 132 \ W/m^2.$$

Compare this to another site where the wind is 3 m/s one-half the time ($\tfrac{1}{2} \times 1.22 \times 3 \times 3 \times 3 = 16 \ W/m^2$) and 9 m/s the other half the time ($\tfrac{1}{2} \times 1.22 \times 9 \times 9 \times 9 = 445 \ W/m^2$). The mean power density at this site is $(16 + 445) \times \tfrac{1}{2} = 231 \ W/m^2$. In both cases the mean speed was exactly 6 m/s (13.4 mph), but there is 75% more energy at the site with varying wind speeds.

At typical sites, the wind speed varies constantly, so the mean wind speed is a poor indicator of available energy. At Point Arena, California, where the average wind speed is 6.7 m/s (15 mph), the mean power density is 421 W/m^2 or 2.3 times greater than the power based on the mean wind speed ($\tfrac{1}{2} \times 1.22 \times 6.7 \times 6.7 \times 6.7 = 183 \ W/m^2$).

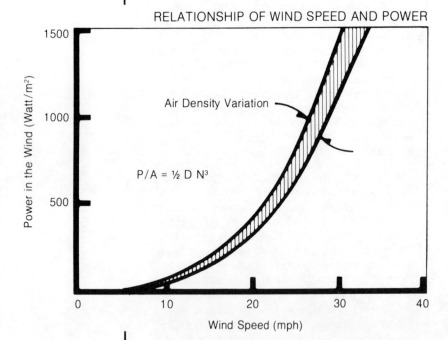

RELATIONSHIP OF WIND SPEED AND POWER

PUMPING WATER

The famous old farm water pumper is discussed here first. This allows us a chance to develop some concepts we'll need later to figure out which type of windmill is appropriate to any given situation.

The farm water pumper functions by driving a piston in a water pump up and down. This requires a driving rod — called a **sucker rod** — being driven by a crankshaft, which is in turn connected to the multi-bladed rotor. Farm water pumpers bring nostalgic memories with their many blades turning slowly in a breeze, but they are actually well suited to the task of lifting water from deep in the ground. The water in well pipes is a heavy load on the piston pump, when it is just starting to move. This means that the windmill rotor blades must supply tremendous starting torque to coax the whole works to begin pumping.

If you stand out in front of one of these windmills, you quickly notice that the rotor blades nearly block your view through the rotor. They form a nearly solid frontal area to the wind. This **solid frontal area** or **high solidity rotor** is the key to the farm water pumper's long success in lifting water directly out of deep wells. By way of immediate contrast, **windchargers** or **wind electric generators** have much thinner rotor blades, more like airplane propellers. Wind electric generators don't have to start spinning under heavy load, so less rotor solidity is needed. The diagram illustrates how the rotors' proportions change according to **rotor speed,** or **rpm,** and solidity needed, which is determined by the task to be performed.

The water pumper is a slow-turning rotor. It produces high starting torque and reasonable efficiency in its slow-turning task. Try to speed up a water pumper's rotor rpms, however, and efficiency will fall off. Because of this, you can't expect a water pumper to make a very good wind generator.

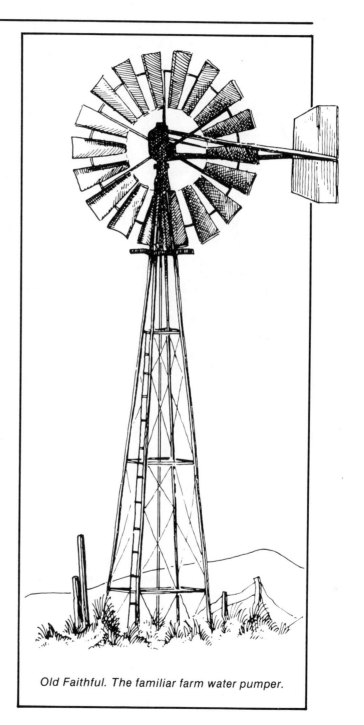

Old Faithful. The familiar farm water pumper.

The familiar slab tail-vane on the water pumper serves two very useful tasks. First, it keeps the machine pointed into the wind when it is running, and second, it keeps the rotor pointed out of the wind during very high winds, when the machine is shut off because water is not needed or during maintenance. Some form of governor on a wind machine is always needed to prevent overspeeding in high winds; the tail-vane does this for the water pumper.

Because water pumpers have been around for so many years, users have had the opportunity to solve many of the problems caused by environmental hazards. For example, water pumpers in the desert used to be installed on wooden towers. But cows visiting the windmill to drink from the stock pond often would indulge in the pleasures of chewing on the tower. This eventually resulted in a collapsed windmill and thirsty cows. Today, steel towers withstand such environmental hazards.

Over the years, people have also found different ways to use wind power to pump water. One is to generate electricity with a wind generator, and then use the power either directly or in concert with utility power to run a well pump. You have the choice of generating either DC (**direct current**) electricity, similar to what is in your car, or AC (**alternating current**) electricity much like what is available from your electric company.

If you choose DC, you drastically reduce your options for directly coupling to a well pump; not many DC well pumps are available. Direct current can, however, be converted to alternating current by means of an electronic device called an **inverter**. These devices are available in rated capacities that range from a few watts to hundreds of kilowatts. Each horsepower of rated pump capacity will need nearly one kilowatt of inverter capacity, plus a surge capability to withstand the increased needs of the pump as it starts up. If you install an AC wind machine that is capable of being wired directly to your power lines, you are directly driving your well pump from the wind generator, with the utility helping as needed. If extra power is available from the windmill, the excess is fed into the power lines to be used by a neighbor.

A third way to lift water is to drive an electric well pump directly with a wind machine without

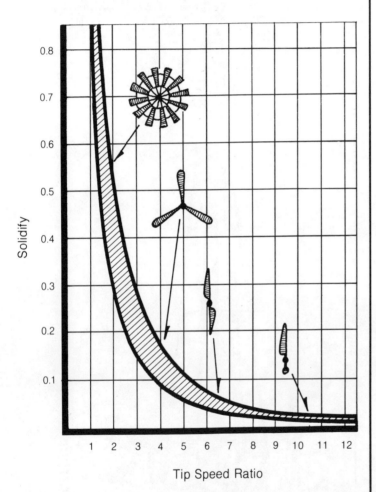

Rotor blade requirements. More blades make more torque for heavy mechanical jobs like water pumping. Electric generators require less torque, and fewer blades.

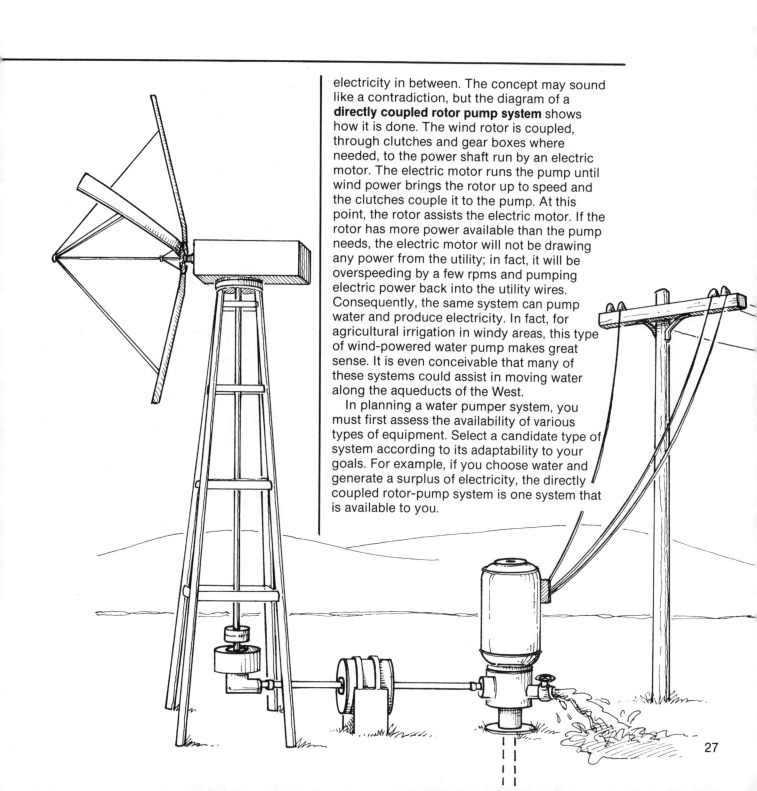

electricity in between. The concept may sound like a contradiction, but the diagram of a **directly coupled rotor pump system** shows how it is done. The wind rotor is coupled, through clutches and gear boxes where needed, to the power shaft run by an electric motor. The electric motor runs the pump until wind power brings the rotor up to speed and the clutches couple it to the pump. At this point, the rotor assists the electric motor. If the rotor has more power available than the pump needs, the electric motor will not be drawing any power from the utility; in fact, it will be overspeeding by a few rpms and pumping electric power back into the utility wires. Consequently, the same system can pump water and produce electricity. In fact, for agricultural irrigation in windy areas, this type of wind-powered water pump makes great sense. It is even conceivable that many of these systems could assist in moving water along the aqueducts of the West.

In planning a water pumper system, you must first assess the availability of various types of equipment. Select a candidate type of system according to its adaptability to your goals. For example, if you choose water and generate a surplus of electricity, the directly coupled rotor-pump system is one system that is available to you.

Once a system type is selected, various aspects of its installation and ownership must be evaluated. How much room is needed to get in installation equipment? Will cranes and backhoes be necessary? How much ground is needed for foundation, guy wires, cables, shafts, pipes, or other equipment? These and other questions must be answered. Further estimates should be made about the type, frequency, and cost of scheduled maintenance routines recommended by the manufacturer. When all of the questions are answered to your satisfaction, determine the exact size and specifications for the system. How to make that determination is discussed further in Chapter 6.

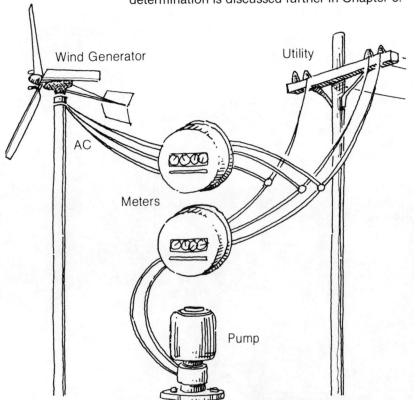

Wind electric water pumping. Electric power is provided by the wind system in parallel with the grid.

PRODUCING PROCESS HEAT

Wind energy can create heat by either of two methods: **mechanical energy transfer** or **electrical energy transfer**.

Vigorously shake a bottle of water, and you will raise the water's temperature. Although the temperature change in this small experiment may be difficult to measure without sensitive instruments, you are transferring energy to the water through mechanical action. This process has been tested on a larger scale with windmills.

A spinning windmill rotor can create heat by either of two mechanical methods: splashing or pumping. Splashing is done by coupling a power shaft from the windmill rotor to paddles immersed in a tank of water. As the paddles spin, water is thrashed around, and the energy the paddles give to the water must be dissipated. Because of friction, that energy ends up as heat. The second method — pumping — has been used with both water and air. A water pump can be driven by a windmill so that water is pumped within the confines of an insulated tank. This water is simply pumped out of the tank and then back into it. The splashing that results causes friction, chaos, and heat. Similarly, an air compressor can be driven by the windmill so that heat energy can be added to large volumes of air.

One other mechanical method for creating process heat is to drive a heat pump's compressor directly with a windmill rotor. This scheme allows for the generation of both high temperatures for heating and low temperatures for refrigeration.

The primary considerations for generating mechanical process heat are based on the unpredictability of wind energy. A relatively expensive investment in a wind-powered heat pump, for example, may not be justified unless that which uses the heat is capable of surviving without the wind. Several sources of back-up heat energy are available. If the heat is used for climate control in a home, other sources include solar energy, wood, and natural gas or electricity. If the process is used in a business or industry, the usual back-up heat sources are probably already in place.

It may be more economical to generate heat from wind energy together with conventional heat-producing techniques rather than by trying to "go it alone." Electric heat energy transfer offers such an option. Simply generating electricity at a windmill offers the opportunity to create heat directly using any of several means: electric heater probes immersed in a fluid tank or air duct, driving an electric heat pump, or simply adding to utility electricity already being used for heating.

Electric heat generation is a technology that is readily available commercially. The techniques of using splashing, pumping, or direct-drive heat pumps are still in development and are only available to energy consumers as do-it-yourself or experimental projects.

GENERATING ELECTRICITY

Along with water pumping, generating electricity has been the traditional use for wind energy at the individual, energy-consumer level. Wind systems that charge batteries have historically been the main type of system used. However, since electric utility lines extend to nearly all users (homes and commercial enterprises), battery-charging wind systems have usually been used in out-of-the-way places such as mountainous regions.

For those consumers not connected to a utility power line, wind-powered battery systems have a place. They are also useful for supplying power to remote electronic instruments, microwave repeater stations, and forest service or fire lookout stations. Determining load characteristics, wind

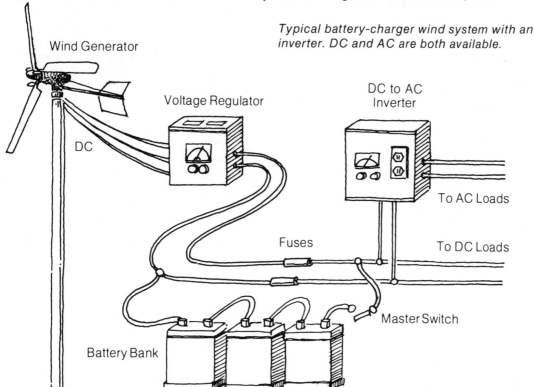

Typical battery-charger wind system with an inverter. DC and AC are both available.

characteristics, and wind system performance becomes very important with battery systems. Unless a back-up source of energy is always available, the system must be carefully planned to prevent complete discharge of the batteries. Battery life and wind generator performance are reduced when batteries are allowed to be fully discharged during windless periods.

Today most appliances require either 120-volt or 240-volt alternating current electricity at a frequency of sixty cycles per second. Since all batteries store DC electricity, these systems require inverters.

A logical method of supplying AC electric power for your appliances is to produce the power in cooperation with the utility company. This has the very distinct advantage of avoiding a fairly significant investment in batteries which are needed for the battery systems. Additionally, no extra wiring or other additions to the system are necessary to make the electric power suitable for appliance use.

Generators that produce alternating current and that connect directly to the utility power lines are rapidly becoming the standard type of system being considered by today's windmill buyers. The arrangement not only lets an energy consumer become an energy producer in parallel with the utility's grid, it also offers the advantage of working directly with existing sources of energy and the appliances that are already in place. For most U.S. installations, this type of system — called a **grid-connected** or **grid-parallel system** — may prove to be the most cost-effective choice available after a careful examination of alternatives. By operating in parallel with the grid's electricity, you don't have to consider exactly what load will be powered by wind. Simply determine the percentage of the entire load you hope to power, and select a wind system capable of matching that goal with the wind energy available at your site.

BEING YOUR OWN UTILITY

A federal law passed in 1978 establishes operating conditions affecting both utility companies and wind power producers who wish to interconnect with the utilities' lines. The Public Utility Regulatory Policies Act (PURPA), has three major elements. First, PURPA requires regulated electric utility companies to interconnect their power lines with small power

Utility interconnect: one method for households. The upper meter measures energy produced by the wind machine. The lower meter measures energy consumed in the residence.

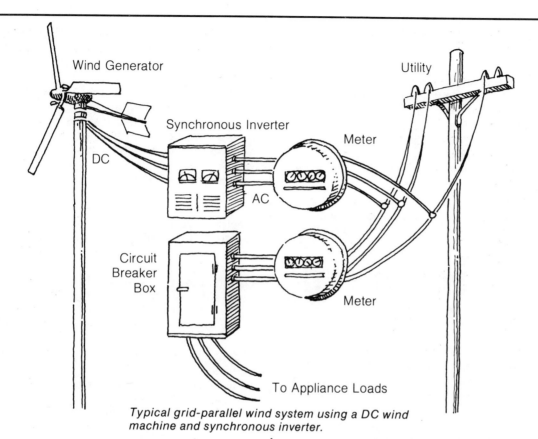

Typical grid-parallel wind system using a DC wind machine and synchronous inverter.

producers who request interconnection. Second, PURPA mandates that a minimum rate be paid to small power producers for the electric energy they feed into power company lines. Finally, PURPA's exemption of small power producers from laws which regulate large-scale power producers — utilities — creates the circumstances making the first two events possible.

The diagrams show how grid-parallel systems can be connected. These arrangements not only let you, under PURPA, become an energy producer or "mini-utility," they offer the advantage of working directly with existing energy sources and the appliances designed to use AC current.

With these set-ups you can achieve a real measure of energy self-sufficiency without cutting yourself off from the grid. Furthermore, by interconnecting with your utility, any excess power you produce can be sold outright or bartered to substantially reduce or effectively eliminate your electrical energy costs.

PURPA, in its insistence on avoided costs, provides a tremendous incentive for people willing to become electric power exporters. It is conceivable that large land owners (such as farmers) in windy areas could tap their wind resource with a number of generators, and set up **wind farms** alongside their other uses of the land.

Despite the fact that PURPA became federal law in 1978, most state regulations for compliance with the law are, as of this writing, still being finalized. The potential impact of the law meanwhile, is so far-reaching that its full effect will not be felt for some time. In the interim, several difficult issues must still be resolved.

Should you choose to link your wind system to your utility, the utility is required to purchase power from you at rates up to **avoided cost**. Avoided cost is what the utility company would have paid had it purchased the power it buys from you from another source or built new generating facilities to supply this additional power.

The electricity you purchase today is generated by a mix of old and new power plants, and the rate you pay is based on the average cost of production. With energy costs increasing, power from newly built plants costs considerably more than that from the current mix of new and old facilities. As a result, avoided costs will typically be higher than the rate you currently pay.

Some exceptions to this rule exist where **inverted block schedules** have been instituted. This arrangement of rates creates two or more tiers of prices, so that users of large amounts of electricity are charged at a higher rate than average prices, while people using small amounts of electricity are charged less than the average. As a result, prices in the upper tiers may be higher than the avoided cost rate. If you are paying these high rates, you will want to negotiate an interconnection agreement to offset your own electrical use rather than selling all your power to the utility at its avoided cost rate.

Other factors will undoubtedly influence your plans for utility interconnection. It may be, for example, that the additional cost of the metering equipment you will need to install if you sell the utility all the electricity your wind generator produces will outweigh the revenues it might generate; with a smaller system this interconnection arrangement will not be advantageous. Your utility, meanwhile, will also be seeking the best terms, and some will only allow certain kinds of interconnections.

Economics aside, whatever the final result of your negotiations, the utility will require your close cooperation on several practical matters of interconnection.

The primary issue is safety. Utility company technicians, servicing the lines, must be protected against electric shock from unexpected power surges, for example, coming from your wind system. Your utility will want to be reassured that your wind generation system cannot endanger its personnel.

The next important issue is the quality of electricity your wind system generates. Electric wind systems, as mentioned before, produce either direct or alternating current. If the grid-parallel system produces direct current, the utility will require a **synchronous inverter** to convert the direct current into alternating current at voltages, frequency, and phase appropriate to the utility's power lines. The particular inverter you select must be acceptable to your utility's engineering office. Consult with them and with your wind system dealer before selecting your equipment.

An alternative method for coupling a wind system to the grid is with an AC wind generator designed especially for this purpose. Selection of such a wind system eliminates the need for a synchronous inverter altogether. Yet even in purchasing an AC generator, you must still consult with the local utility and wind system dealer to determine whether your candidate wind machine is suitable for an interconnection.

SELECTING EQUIPMENT

In selecting equipment for a wind energy system, you must consider the job you want it to do, your budget limitations, the upkeep it requires, and the social and institutional issues. This process is similar in many respects to the decision-making process used when buying a car.

The type of car you purchase must do the job of getting you from one place to another. Most people complicate the process by also

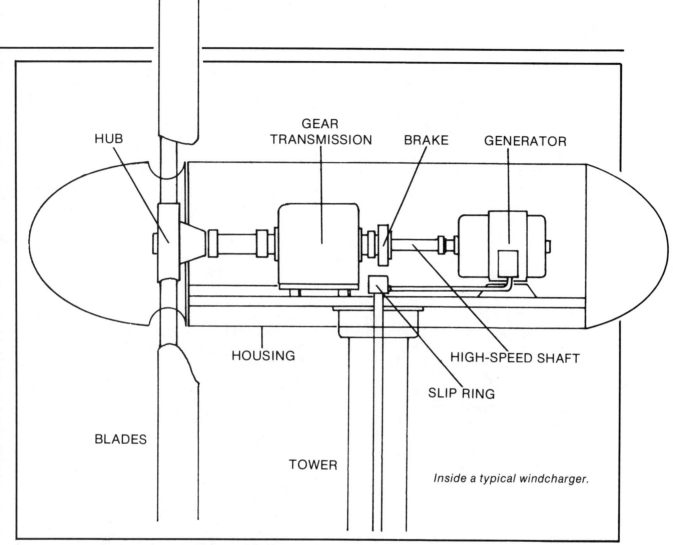

Inside a typical windcharger.

requiring that it get them about in a certain style and measure of fuel economy. A wind energy system's job is to supply energy in the form of heating, water pumping, or electricity. This is also complicated by the importance of matching the wind system to the site characteristics and the energy load. Energy needs and methods of assessing these needs were discussed in Chapter 2. Matching the wind system with site characteristics is discussed in the next chapter. Social and institutional issues are considered in Chapter 5.

When you shop for a car, you are likely to look very hard at the warranty, who the manufacturer is, the maintenance and repair records of the model, as well as cost of repair, availability and cost of spare parts, and the number of shops and mechanics who can work on it. Unfortunately, comparison shopping is very difficult with wind energy systems. Unlike the auto industry, the infrastructure of marketing, supporting services, spare parts, and repair for wind systems is only now beginning to develop.

First recognize that wind energy is an industry in its infancy. This has several important implications. There are few government regulations providing for consumer protection and public safety and many installations are subject to review by local building and electrical inspectors who do not understand the equipment. Also, there are new companies and products being introduced as the market expands. Some of these may not be in business five years later. Finally, consistent with the American tradition of free enterprise, there are always some sales and promotion companies, who, knowingly or otherwise, tend to exaggerate what their products will do. All of these suggest that you, as the potential buyer, must be as informed as possible.

Selection of a wind machine is usually dictated by the machine/load/site match. For example, if your energy load is for pumping water for irrigated pasture, there is not much sense in comparing the aesthetic appeal of a high-speed electric generator with a Darrieus machine. Other considerations, like service agreements and availability of parts, will usually sway a decision about machine selection more than a comparison of peak efficiency.

There are two types of towers used for wind installations: **guyed** and **free-standing**. Guyed towers tend to be lower in cost but you must have space to secure the three anchor points. Free-standing towers take up less space on the site, but require a fairly substantial chunk of concrete to anchor the whole system to the ground. If aethetics and space are chief considerations, then a free-standing tower is a good choice.

A wind machine on top of a tower is a dynamic system that vibrates in the wind. Designers try to select equipment that minimizes major vibrations. For this and other reasons, most manufacturers specify their choice of tower for their machines.

It is a good idea to plan your electrical wiring ahead of time and have an electrical inspection done for your installation, even if your county doesn't require it. An experienced electrician can usually spot deficiencies and offer suggestions during a short visit. When planning your system, be sure to consult your local electrical codes.

Grid-connected wind systems require prior notice and approval of the utility company. If you plan on doing this yourself, you should have plenty of experience doing electrical wiring. Otherwise, hire an electrician to assist with the installation.

Even at very windy sites, the energy in the wind isn't always available when you need it. So for non-grid-connected electric systems, storage batteries are required. Lead-acid batteries still represent the cheapest practical method of electrical energy storage. Storage batteries recommended for wind systems are marine or deep cycle batteries. They are usually much larger and have thicker lead plates than car batteries for repeated cycling over many years. They can be discharged and recharged many times.

As a rule of thumb, battery storage capacity should be large enough to carry your normal load for at least three days without energy-producing winds. If electricity is not that important to you, you can do with less battery storage. On the other hand, if you are running a freezer and don't want to be without power, you may opt for larger battery storage and a back-up power source. Here is a case where you must balance the cost of more batteries against the escalating fuel cost of a back-up generator.

Back-up power sources are usually used with off-grid, battery charging wind systems. The back-up system, of course, is only called on when the batteries are discharged, but it should be sized to accommodate your peak load. This requires a certain amount of discretion. You probably don't want to run your electric water heater on a gasoline powered back-up system, for instance. Back-up generator systems are widely available in all sizes, so with this equipment comparative price shopping is advisable.

Batteries have been used as electrical storage for wind systems for several generations, so information on their proper care is generally available. However, a few pieces of free advice still bear repeating: keep your batteries inside in a well-ventilated space, off the floor and away from children and animals.

If you store your electricity in batteries, how do you use it? A lot of appliances — such as heaters, lights, and your electric drill — don't care whether their power is AC or DC. Stereo and television sets, however, only use sixty cycle AC power. To plug them in, a DC-to-AC inverter is necessary.

When selecting these inverters, you need to know about quality, efficiency, and power rating. Power quality refers to the kind of AC signal produced. Your television set uses a **sine wave** signal and nothing else will do. Some inverters produce a **square wave** signal. The old Army surplus rotary inverters put out a very high quality sine wave signal, but are only about 60% efficient. Solid-state units are about 80% efficient. Inverters are rated as to their maximum power capability. Remember that a 5000-watt inverter is only capable of converting 5000 watts of power; it won't produce power for you. Size the inverter according to the maximum you expect to demand.

The other type of inverter is the synchronous inverter mentioned previously. This device goes between your windmill and the powerline for connecting to the grid. It changes the DC electricity (or varying frequency AC if you have a non-rectified alternator) to an AC signal synchronized with the power company's signal. Synchronous inverters are relatively new on the market and your selection may be limited to an approved list provided by your power company. Synchronous inverters can be noisy and like batteries require free air cooling; put yours in a safe place.

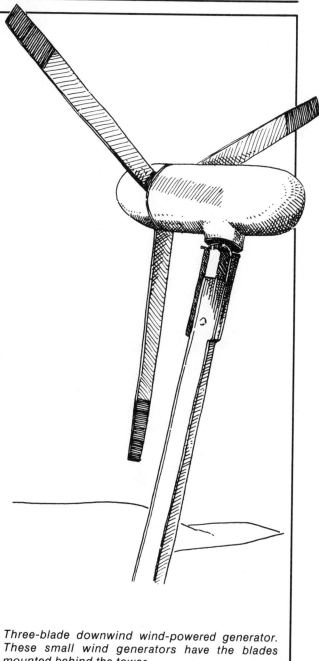

Three-blade downwind wind-powered generator. These small wind generators have the blades mounted behind the tower.

Chapter 4

FUNDAMENTALS OF SITING

Planning a good wind system requires paying close attention to several key issues. Previous chapters have discussed some of them—your energy needs and design based on your goals for the system. Some new issues also need to be covered. Two important points you must consider in selecting a good site are:

Ensuring safety
- Keeping machine separated from neighbors' yards and buildings.
- Preparing for seismic disturbances in appropriate areas.
- Preparing for hazards of extreme winds.

Maximizing energy production
- Finding the place where wind energy is best.
- Determining how high the tower should be.

ENSURING SAFETY

Safety is an important issue in siting wind machines. If you have decided that enough wind energy is available across your site and that you want to plan an installation, the first issue to deal with is safety. A wind machine is an attractive nuisance in much the same way a swimming pool is. A windmill tower is as interesting as climbing a jungle gym at the local park, but is vastly more dangerous, especially if the blades at the top are spinning. All sorts of people will be attracted to the tower, so it should be sited in a safe area and surrounded with a safety fence.

Somewhat akin to that issue is the hazard created by the wind machine simply because of its presence above people and property. As a general rule, the machine should be sited far enough from people and property that, should it topple, it cannot strike and damage houses or the people inside. The siting process allows some compromise in selecting a place to dig the foundation hole and pour concrete to support the tower. No compromise should be made, however, in relation to these two safety issues.

Seismic safety is also important in various parts of the U.S. and particularly in California. When planning a wind machine foundation, do not overlook the potential for damage to the foundation during an earthquake and for damage to the wind system's structure from other objects moving during a quake. A landslide or boulders tumbling down a hill near a tower could damage both the tower and the wind machine itself. You must anticipate such hazards.

Extreme winds must also be considered. Once every ten to twenty years, the wind is liable to blow as fast as 100 mph across your region. Site analysis includes looking for data that will give some clue to how fast you can expect an extreme wind to blow. A nearby weather station may have figures for both the average wind speed and also for the fastest one-minute gust ever recorded there. Don't overlook extreme wind just because it happened a long time ago, either. It can always happen again, and your system should be able to withstand a wind so big it only blows that hard once in a hundred years.

MAXIMIZING ENERGY PRODUCTION

Before investing in wind energy equipment it's necessary to evaluate the wind energy potential of the site (or sites) you have in mind for your future windmill(s). There are two approaches to wind assessment; the one you choose will depend primarily on the amount of land area involved.

If you go looking at sizeable areas for valuable wind energy potential, you are wind prospecting, recalling days when the '49ers, seeking gold, tramped about the Sierra foothills. If you have a particular site in a specific location that you wish to assess (say under fifty acres), you are involved in site analysis.

Wind prospecting yields general information on the wind energy available in a region, though it may include valuable detailed data on the quality of the wind resource for specific sites monitored by wind energy measurement equipment such as **anemometers** (wind speed sensors).

Site analysis yields specific information on the wind energy available at a single location only. It can provide the data you need to make a careful estimate on wind machine performance should you "plant" one at the site.

In a wind prospecting project, individuals or organizations start by collecting all available wind data already gathered by meteorological stations or previous wind energy research. Some of this data is summarized in Appendix 3. The data is then plotted on a map, and coded to represent the amount of available energy. This wind energy prospecting map is then overlaid on another map displaying urban areas or other locales with the greatest energy needs. Electric power lines are also drawn in so that windy sites nearest to available transmission lines may be reviewed first for priority selection. Only after this initial work has been completed are anemometers and recording instruments installed. As a result of this type of ambitious program, areas suspected to possess only marginal wind energy potential have on occasion surprised prospectors, much to their delight.

After collecting regional data, the wind prospector may well undertake site analysis to confirm how much of the prospected region falls into the three different categories listed below:

- obviously windy enough to pursue a wind project;
- not sure;
- obviously not windy enough to pursue a wind project.

If you can determine that your site's annual wind speed average is at least twelve mph, you probably have found an "obviously windy enough" site. This criterion is, however, somewhat arbitrary, and you should consult your wind system dealer for his or her own recommendations. Whatever evaluative criteria you settle on will profoundly affect your decision on what wind machine to purchase.

Typical "cup" anemometer. Wind sensors like this one are used for site analysis and prospecting.

Most sites will fall into the "not sure" category, but many will fall disappointingly flat, ending in the "obviously not windy enough" category.

If a small wind system is planned, those sites falling into the "windy enough" category may require little or no site analysis. But when larger wind systems — fifty kW or above — are contemplated the cost of installation alone certainly justifies the expense site-specific wind data collection will entail.

"Not sure" sites will require more in-depth research beginning with a search for **proxy evidence** of wind energy. Proxy evidence may be found in flagged trees, sand piles organized in uniform patterns, or bent brush. This evidence, if present, will help you decide whether to proceed with more detailed evaluations. Strong proxy evidence is an excellent indication of the wind's presence. If you find this evidence, proceed with site analysis.

The next step should be to purchase, rent or borrow site evaluation instruments. You should consult with your wind system dealer for prices and availability. The cost of a typical site analysis is under $1,000—far less then the cost of most small wind system installations.

A few months of instrumented data collection will help you determine whether a "not sure" site is windy enough, or falls out of the running. About one year of data collection is necessary to give truly accurate estimates of the economic worth of your site's wind power.

Whether you are prospecting or doing a site analysis, wind energy is essentially what you will want to establish. This may include a number of elements such as **mean power, mean energy wind speed** (which is discussed below), and the mean or average wind speed.

The illustration shows a typical anemometer installation, much like an installation you might use to assess your site's wind energy. This instrument is mounted atop a support mast—usually thirty feet above the ground and well above the tree tops—with a signal wire leading down to a recorder. A **strip-chart recorder** is used, but this device is being replaced by **solid-state recording devices**. A close examination of the strip-chart record shows that the true wind speed is recorded at every instant. From this recorder it is possible to obtain a wind speed record corresponding to regular time increments, such as once every hour.

From the history of wind speeds in the strip-chart illustration, it is possible to count up the number of times each wind speed occurs. The bar graph shows the number of times each wind speed (in one-mph increments) was observed over the period monitored (one month, for example). The graph shows that wind speeds are more frequent at ten mph than at five mph or at twenty mph.

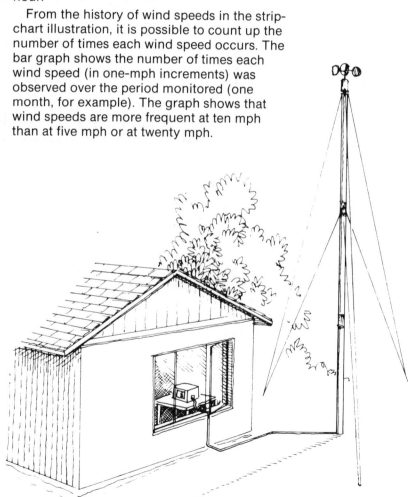

Typical anemometer installation. The wind sensor (anemometer) is mounted well above the trees.

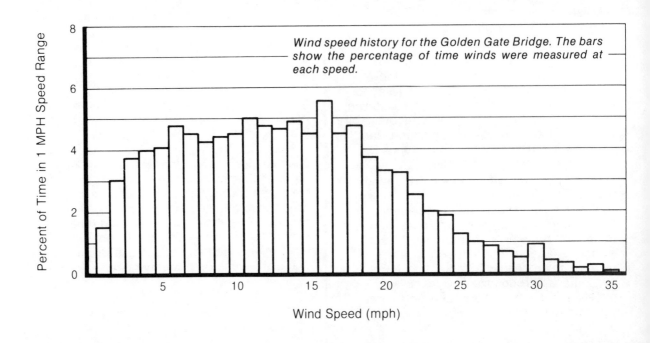

Wind speed history for the Golden Gate Bridge. The bars show the percentage of time winds were measured at each speed.

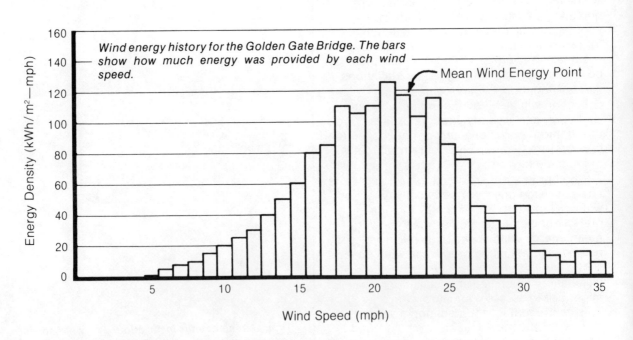

Wind energy history for the Golden Gate Bridge. The bars show how much energy was provided by each wind speed.

Mean Wind Energy Point

As wind speeds increase over ten mph, their frequency tends to decrease. While the most frequent speeds measured at the majority of U.S. sites will be in the five to twelve mph range, the graph verifies most people's intuitive sense that the wind blows regularly at low speeds and less at high speeds. A bar graph such as this can be generated for each site using a strip-chart recorder but the graph is not needed for a complete site analysis. Data, when collected specifically to analyze wind energy, renders such a graph redundant.

Wind power or wind energy is the primary element in which you should be interested. The power available in the wind is directly related to the cube of the wind speed. That is, if the wind speed is doubled, eight times the power is available.

Knowing this, it is possible to convert the bar graph of wind speed occurrences into a similar graph illustrating a site's available wind energy. This is by far the most important graph that can be plotted for a given site's resource. The graph is created by multiplying each wind frequency (taken from charts such as the wind speed occurrence bar chart) by each wind speed cubed.

Such a reconstruction is illustrated in our diagram. The energy distribution in this diagram shows that most of the energy in the wind is available in the range around twenty mph. This makes it apparent that the wind speed corresponding to the weighted center of the energy distribution is a very important measure of where wind energy occurs. This mean energy speed is the balance point, as if the curve were cut out and balanced on a fingertip.

Mean energy wind speed tells you the wind speed above which nearly half of your energy is available, and below which just about the other half is available. Mean energy speed suggests the speed range where we can expect to get most of the energy in the wind.

As it turns out, the mean energy wind speed for a given site is between about one and one-half and three times the average wind speed. Thus, if the average wind speed is measured at, say ten mph, mean energy wind speed might be, say twenty mph; but its value cannot be guessed—it must be measured or carefully estimated with formulas using data measured at the site. Once you know such a speed, you can easily select wind machines whose performance centers in the same speed range.

The next logical question is: how much power is available in the wind across a site? Mean power available is typically presented in units of **watts per square meter** (square meters of windmill frontal area—which can easily be converted to square feet).

The power available in the wind is proportional to the cube of the wind speed. That fact has two extremely important effects on data requirements for wind site analysis (or wind prospecting, for that matter). First, the cubic relationship between wind speed and power suggests that estimates of available power are very sensitive to speed measurements. It follows that inaccurate speed measurements can introduce large errors in energy estimates. For example, if your anemometer has a 3% error, you end up with nearly a 10% error in power and energy estimates.

Second, the wind itself is so variable. Wind rarely blows for very long at a given speed; it is always changing speed. To see how this variability of wind speed affects your power estimates, consider two hypothetical sites having average speeds of exactly fifteen mph. At the first site, the wind is always constant at fifteen mph—therefore, the mean power available is 191 watts per square meter. At the second site, the wind is ten mph half the time and twenty mph the other half. The mean power available at the second site is 255 watts per square meter, or 33% greater. This example points out that variability in the wind makes the mean power available greater than one would estimate on the basis of the average wind speed. Generally, the greater the variability, the greater the difference between energy estimates based on mean speeds and

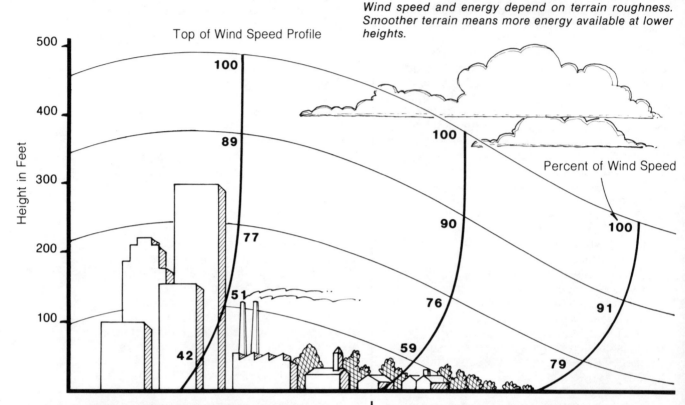

Wind speed and energy depend on terrain roughness. Smoother terrain means more energy available at lower heights.

measures of the true available energy. Whenever possible, instruments designed to establish both available energy and mean energy wind speed should be installed. Calculations of the economic benefit of your site's wind energy are profoundly affected by the accuracy of your measurements.

Remember, too, that while you may measure wind energy at a height of, say thirty feet above the ground or tree tops, you might want to install your wind machine several feet higher. Wind energy increases with height, so each foot of increased height yields increased energy. The diagram illustrates how wind speed increases with height for different types of terrain. If your wind energy comes from across smooth terrain, such as across a lake, you don't have to reach as high for higher wind speeds as you would in rough terrain such as forests or towns.

This terrain or surface roughness effect on wind energy suggests that you select a site that has the least interference to wind flow. It also suggests that the highest possible tower is best. However, the formulas presented in Guide 3 show that wind power available from your machine is not only proportional to the cube of wind speed, but is also directly proportional to the frontal area of the rotor—which increases with the square of the **rotor diameter** (double the diameter equals four times the frontal area).

All of these mathematical relationships tell you to strike a balance between the height of the tower and the diameter of the wind rotor. The balance, obviously, will be your budget for the equipment. An experienced wind system dealer or consultant can help you assess the effect of various tower heights on power available, but you can easily see that if your wind flows from a smooth direction, tower height will be less important; so larger diameter rotors can then be considered. In rough terrain, however, you must strive to reach up into the more smoothly flowing wind stream for your energy; trees, buildings, and other ground level roughness really sap power from lower level winds. Owners' manuals for many wind machines suggest that you plan to install your wind machine twenty or more feet above the tallest trees around.

DOING YOUR OWN WIND STUDY

You should follow a carefully planned procedure in performing your own wind site analysis. The steps you might take are outlined here:

1. Select sites on your property according to safety, noise, appearance, seismic, and clear-ground criteria.
2. Rank those sites according to obstructions to wind flow — if you only select one site, ranking is quite easy.
3. Collect all available information on wind energy at or near your site. Data from more than a mile away, in rough terrain, or ten or so miles away in smooth terrain should not be regarded as indicative of your site's energy potential.
4. Search for any proxy evidence of wind activity. Evidence found in flagged shrubs and trees may help you determine from which direction your wind energy comes.
5. Either make a decision based on existing data and proxy evidence to install a wind machine, and forego more accurate performance predictions or put instruments at your highest-ranked site.
6. If possible, plan to leave your anemometer and other instrumentation installed for one year or longer. A minimum time would be at least one month from each of two seasons — the most and the least windy. As long as an instrument is needed for two separate seasons, it might as well remain installed for a year. You may, however, make your final calculations and decision before the year is up.
7. Follow the procedures given in the user's manual supplied with the instruments you select. Guide 4 in this chapter suggests a procedure you might follow, in the absence of a user's manual, to arrive at a reasonable estimate of performance for various types of wind machines you may select.
8. Plan to share the data you collect and calculations you make with your wind system dealer, the wind energy experts at your state energy office, and with your neighbors. Sharing will help others to become familiar with wind energy available in your area and perhaps encourage them to install wind systems.

GUIDE 4

ENERGY ESTIMATIONS FOR WIND SYSTEMS

There are many different ways to estimate the annual energy output from a windmill. Whatever method you select, you will always need some information about the wind and some more information about the windmill.

Two methods of estimating the annual energy output from a windmill are presented here. The first method is short, simple, and gives only rough estimates of annual output. This method is appropriate when you know very little about your wind resource and wind machine. The second method, called "the method of bins," is more rigorous and is suitable where extensive wind measurements have been performed. It is likely that a wind consultant would use the method of bins to verify any preliminary assessment.

The short estimation procedure outlined here requires that you know the diameter of your windmill and the mean wind speed at your site.

The diameter of a windmill is the distance from tip to tip of the blades. The diameter is proportional to the swept area, which is the area facing the flow of the wind from which the machine collects its energy. The larger the swept area covered, the higher the amount of captured energy.

If you are considering a vertical-axis wind machine, then the blade diameter is not what you want to use for this exercise. Instead, you need to find the equivalent diameter, as if its swept area were circular. For the purposes of this estimation, find the equivalent diameter by using the following equation:

$$\text{equivalent diameter} = \text{square root of } \frac{4 \times \text{swept area}}{3.14\,(\pi)}$$

The swept area is usually available from the manufacturer or dealer. This is the number to be used for "diameter" in the graph on the following page.

Using the graph, estimate annual energy production by locating your site mean wind speed on the horizontal line. Move up to the line corresponding to the appropriate rotor diameter. Interpolation between lines may be necessary. The corresponding reading on the left is the estimated annual energy output, in kilowatt-hours per year (kWh/yr).

The numbers in this graph are only estimates based on general assumptions about machine performance and wind behavior. Actual performance may vary depending on specific wind machines and sites. The use of this graph is suitable for situations where little additional information is available about the wind resource or about the performance of the windmill. If you know more detail about the wind characteristics at your site and about your machine, then take a closer look at the method of bins below.

An Example of the Short Method

A farmer is considering the use of a 45-foot diameter wind machine at his site which has an estimated annual mean wind speed of 14 mph. How much energy could he reasonably expect to produce from a wind machine in an average year at his site?

SOLUTION: On the graph, find 14 mph on the horizontal scale and move vertically to the intersection with the 45-foot diameter line. Read your answer on the left-hand scale at a height corresponding to this intersection point. The result is about 73,000 kWh per year.

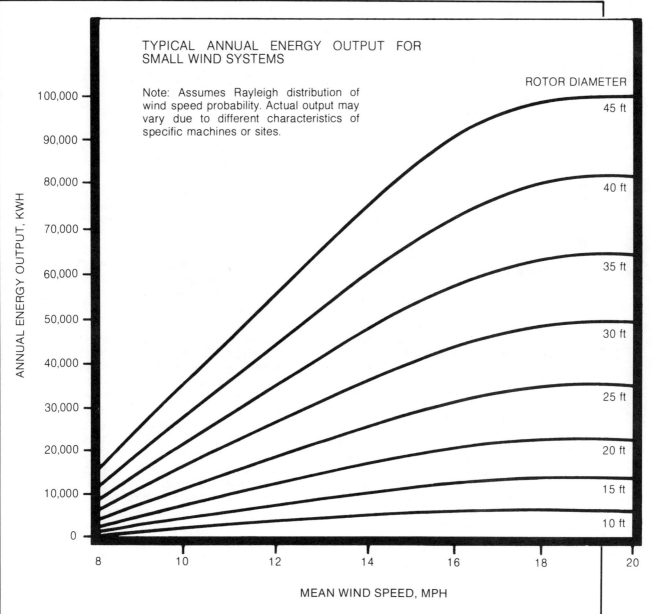

GUIDE 4

Park, Jack and Dick Schwind. *Home Wind Power/Wind Power for Farms, Homes, and Small Industry.* Charlotte, Vermont/Mountain View, California/Washington, D.C.: Garden Way/Nielsen Engineering and Research, Incorporated/U.S. Department of Energy: available also from National Technical Information Service, 1978, 1981.

GUIDE 4

The Method of Bins

The method of bins, presented here, assumes that you know the percent of time the wind blows at each wind speed. This type of information has been presented earlier as a bar chart. Some instruments record this information directly. It can also be determined wherever hourly records of wind speeds have been recorded.

The information you need about your wind machine is the amount of power it will put out at each wind speed. This data should be available from the manufacturer or dealer. In any case, data based on actual performance measurements are likely to be more reliable than theoretical calculations.

An Example of the Method of Bins

The farmer from the last example decides to take measurements for his site for a year before proceeding with his windmill project. His data shows that, for the year of his measurements, the mean wind speed was 13.1 mph and that the mean power density was 237 W/m^2. The frequency distribution from this data is presented in the following table. The percent of time at each wind speed is listed under the "% TIME" column.

The farmer buys a two-bladed, 48-foot diameter, 40-kW induction generator wind machine. After consulting with his dealer, he is able to determine that the machine had been tested according to a wind energy association accepted testing procedure referred to as the method of bins. The information from this test provides him with the power output at each wind speed, as listed in the third column in the following table. What is the estimated annual energy produced by this machine?

SOLUTION: A fairly accurate method of estimating the annual energy generated by a wind machine involves a lot of multiplication and addition of numbers. With the aid of the following example table below, this is actually a fairly straightforward procedure. Perform your estimate by multiplying each "% TIME" by the corresponding "POWER" and write the product in the column called "ENERGY." Now add up all the numbers in the "ENERGY" column.

The sum of the numbers in this column is called the **mean power output**. This is the average power level of the wind machine for the year. In this example, the mean power output is 9.06 kW. Since there are 8,760 hours in a year, the annual energy estimate is the product of the mean power output and 8,760 hours, or 79,374 kWh per year.

It is reasonable to allow for some downtime of the machine due to maintenance and unforeseen repairs. In the example, an 8-percent allowance for downtime is assumed. Thus, the final estimate has been revised downward to 73,000 kWh per year.

In this example, the final result came out very close to the earlier rough estimate of 73,000 kWh per year. In general, such close agreement between the rough estimate and the method of bins procedure would not be expected. Some of the conditions of the estimate were changed. By the time the measurements were completed, the farmer knew that he was using a 48-foot diameter machine, not a 45-foot machine. The site measurements showed a 13.1 mph annual average, not the 14 mph estimated prior to the measurement project.

The most significant difference between the two methods lies in the shape of the frequency distribution curve — the "% TIME" column in the table. The short estimate assumes one common shape of a frequency curve. However, the exact shape of distribution curves varies greatly with individual sites. The method of bins estimates allow the direct use of the actual distribution data.

A continuation of this example is included in Chapter 5 under the discussion on economics of wind installations.

Method of Bins Energy Estimate of a 48 ft. Diameter, 40-kW, Horizontal-Axis Wind Machine

Wind Speed (mph)	% Time	Power (kW)	Energy (kWh/hr)
0	0.50	0	0
1	1.68	0	0
2	3.00	0	0
3	3.66	0	0
4	3.88	0	0
5	3.88	0	0
6	4.59	0	0
7	4.51	0	0
8	4.32	0	0
9	4.34	0.5	0.022
10	4.42	1.8	0.080
11	4.89	3.2	0.156
12	4.67	4.5	0.210
13	4.63	6.3	0.292
14	4.79	8.0	0.383
15	4.56	9.8	0.447
16	5.33	11.5	0.613
17	4.58	13.4	0.614
18	4.79	15.3	0.733
19	3.86	17.1	0.660
20	3.49	19.0	0.663
21	3.43	21.0	0.720
22	2.77	23.0	0.637
23	2.15	25.0	0.538
24	2.07	27.0	0.559
25	1.40	29.0	0.406
26	1.10	31.0	0.341
27	0.61	33.0	0.201
28	0.42	35.0	0.147
29	0.34	36.0	0.122
30	0.42	37.0	0.155
31	0.16	38.0	0.061
32	0.13	39.0	0.051
33	0.07	39.0	0.027
34	0.12	39.0	0.047
35	0.08	40.0	0.032
>35	0.36	40.0	0.144
	100.00 %		9.061 kW

MEAN POWER OUTPUT = 9.061 kW
Estimated Annual Energy = 9.061 kW x 8760 hr = 79,374 kWh
Allowance for downtime: 8%
Annual Energy Estimate = (1 - .08) x 79,374 kWh = 73,000 kWh

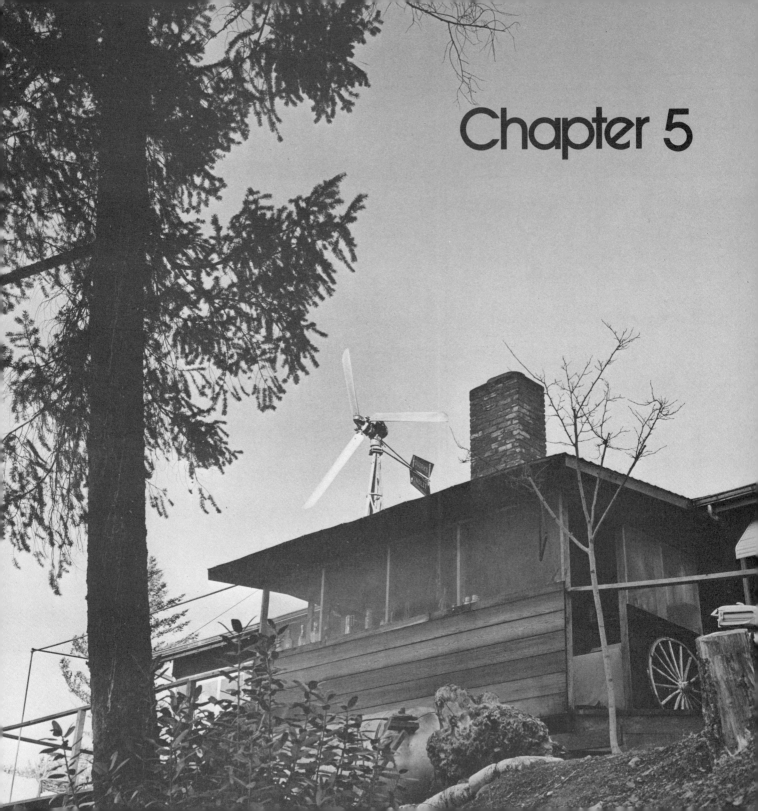

Chapter 5

DEALING WITH ECONOMIC, LEGAL AND SOCIAL ISSUES

Once you have decided to install a wind system, you will have thoroughly analyzed your energy needs and the ability of your wind resource to supply energy, and you will have selected a wind system that appears capable of coping with the hazards of being installed at your site while at the same time turning your wind resource into the energy you need.

But your planning will not have stopped there. You will probably have answered dozens of questions. A typical first question is, "Will the wind power my house?" The appropriate answer is, "Just how fast would you like your house to go?" If you've read this entire book up to here, you know that the answers to the questions you likely had before reading to here aren't all that simple.

By now, however, you've asked, "Can I afford the wind system I want to install?" You might also have asked, and answered, "Will my neighbors object to my windmill?" or "What will happen to my property tax, insurance premium, membership in the Grange, and other such items when I install my windmill?"

You've entered the domain of the hidden issues. Such issues are hidden in just about any purchase you make, but in most cases you have grown used to dealing with them or have learned what results to expect. With wind machines and some solar systems, you might not be ready for these hidden issues. In fact, answers to some do not now exist; new laws are being drawn up and passed which will eventually supply some answers to questions about taxes, liability, and zoning rights for alternate energy systems.

ECONOMICS

Generally, when people spend their money, they are interested in getting the best deal they can find. When comparison shopping, you usually look for some "figure of merit" and compare this figure for several brands of the same product. If you're shopping for a bag of beans, for example, you're likely to compare the number of pounds of beans per dollar. In this case, the figure of merit is "pounds per dollar." With new and different commodities, like windmills, it's sometimes difficult to find a true figure of merit.

People in the windmill business commonly use terms like "Brand X cost me $2,000 per kW." The "dollar-per-kilowatt" figure of merit has a long history in the electric power generation business. But it can be misleading when it comes to windmills. The reason for this is that some manufacturers will attach an oversized generator to a small rotor. Other manufacturers will do just the opposite, depending on the design criteria they try to achieve. In fact, rotor diameter, not generator size, gives a better indication of the wind machine performance.

Since the commodity produced by most windmills is energy, it makes sense to measure the benefit in terms of kilowatt-hours of energy produced in a given year. An easy method of estimating this figure has already been described in Guide 4 of the last chapter. So a figure of merit for smart windmill shoppers would be the number of kilowatt-hours produced annually per dollar of installed cost (kWh/$).

The graph on the next page shows a broad range of this "kWh per dollar of installed cost" figure of merit for a number of wind installations. The kWh/$ ratio is presented as a function of the annual mean wind speed. The trend of a higher figure of merit for higher wind speeds re-emphasizes the importance of proper siting. A good buy on a windmill doesn't mean much if the wind doesn't blow according to your expectations.

Installations considered for this graph included machines as small as 12-foot diameter and as large as 300 feet, but most were under 100 feet. As the wind industry and the technology mature, the cost of installing and producing wind machines is likely to decrease somewhat compared to present costs.

The figure of merit graph gives you a method to quickly compare different ratios of installed cost for your mean wind speed for various types of wind machines. It is intended to assist you in getting started in your decision-making process by providing a quick and easy comparison of different figures of merit. But just as few people buy a car solely on the basis of design or gas mileage, wind machines should not be purchased solely on the basis of overall figure of merit performance. Every machine, site, and power demand is different.

There are many other factors to consider when looking into the economics of owning and operating a wind machine. The rest of this section goes further into the subject of cost accounting for wind installations.

So far, our comparison shopping discussion has only covered the first costs of owning a wind machine. There are other costs and benefits that you could expect to be spread out over a long time. One method of considering the long-term effect of this kind of expenditure is called life-cycle cost analysis.

In life-cycle cost analysis, you not only include the first costs, but add up your best estimate of what the total costs will be over the life of the equipment and compare this to the anticipated benefits.

For some wind installations, this could be longer than 20 years. Two methods of accounting for all the costs and benefits of a sample wind installation are included in Guide 5.

First costs in a typical installation cover the equipment purchase and installation. Installation costs may include the wind siting study, permit applications and fees for your county or local government, trenching, foundation work, assembly of a tower and machine, electrical wiring, and probably a few items you overlooked. Your dealer should be able to assist you with calculating these expenses.

Expenses that cover a long period of time are generally more difficult to figure. These may involve land rental or lease payments, additional property taxes, insurance premiums, and operating and maintenance costs. Many dealers offer warranty service contracts to cover all or part of operating and maintenance costs.

By the time you add up all these costs, things may begin to look fairly bleak for your project. But take heart. Now you need to account for the "up side" of life-cycle costing. The reason for going to all this trouble in the first place was to generate more energy. Most utility companies allow you the option of either displacing your existing energy load or of selling all your power back for direct sales to the company. This translates to dollars saved or dollars earned.

The example in Guide 5 is based on a machine which costs $45,000 installed. This is a 45-ft. diameter machine located on a farm with a 14 mph annual average wind speed. The farmer opted to sell all the electricity generated by the windmill back to the company for 6.5 cents per kWh. In Guide 4, it was estimated that this size machine could produce 73,000 kWh per year. This multiplies out to $4,745 per year of income generated by a $45,000

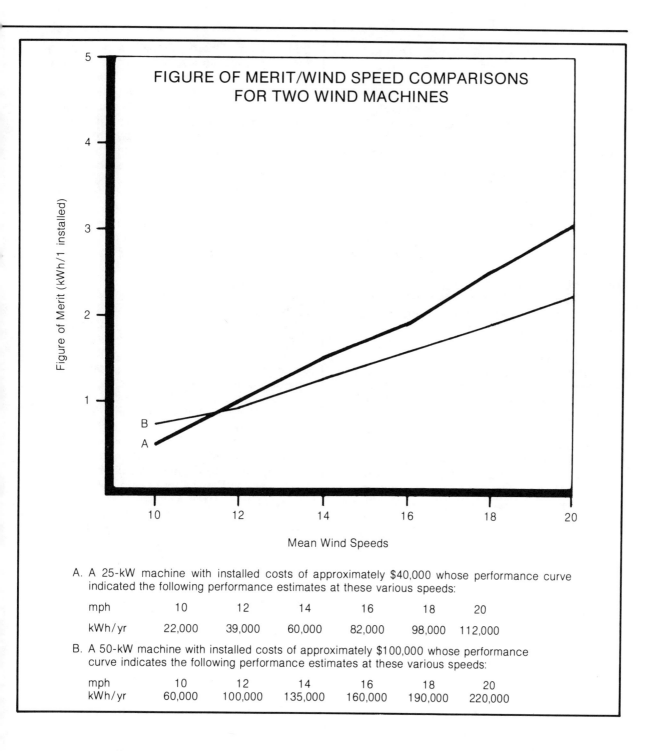

FIGURE OF MERIT/WIND SPEED COMPARISONS FOR TWO WIND MACHINES

A. A 25-kW machine with installed costs of approximately $40,000 whose performance curve indicated the following performance estimates at these various speeds:

mph	10	12	14	16	18	20
kWh/yr	22,000	39,000	60,000	82,000	98,000	112,000

B. A 50-kW machine with installed costs of approximately $100,000 whose performance curve indicates the following performance estimates at these various speeds:

mph	10	12	14	16	18	20
kWh/yr	60,000	100,000	135,000	160,000	190,000	220,000

investment. Without considering any other costs or tax effects, this installation has a simple payback of 9.5 years ($45,000 divided by $4,745 per year).

Up to this point, things have been fairly straightforward. But the labyrinth of government tax computations still has to be dealt with. This book will only scratch the outer surface of a tax subject and, perhaps, send you off to an accountant with a few examples.

Although federal and state tax laws change frequently, a variety of benefits are built into the tax structure which may substantially improve the financial aspects of your project. Most of the available tax incentives only benefit people who pay income taxes. If you don't pay any taxes, you cannot lower them. A tax credit is not to be confused with a tax deduction. Deductions only reduce the amount of your income that is taxable; tax credits are subtracted directly from the amount of tax money you owe.

Further information on this tax credit law can be obtained in IRS publication #903, "Energy Credits for Individuals." The form required to claim the credit is #5695.

The energy tax credit laws for commercial applications can also provide substantial economic incentives for investors. The federal law for commercial applications — the wind installation is part of a business rather than a dwelling — is somewhat more complex. This business energy credit is refundable rather than a carry-over as with the residential credit. This credit can be claimed only after the regular investment tax credit has been claimed.

LEGAL ISSUES

Laws that restrict your rights concerning wind energy fall into categories such as deeded restrictions or covenants, zoning laws, and building codes. Architectural restrictions usually abound in so-called "planned communities" where committees must approve (or disapprove) your request to do just about anything to your property. Zoning laws set aside land in various areas of counties for different purposes. Agricultural, residential, and commercial activities are typical of the domain of zoning laws. Such laws might include maximum structural height limitations, which are usually too low for practical wind energy applications in residential areas. Building codes tend not to be restrictive, but rather prescriptive of how you must go about doing something. Note that building codes might restrict your ability to install a wind system, not because they restrict you from doing so (they don't), but rather because the safety requirements they pose on foundation design, for example, might make the project prohibitively expensive.

In all cases of imposed restriction, you have one final right: the right to protest and gain a variance, or permit to deviate from the rules, according to an agreed-upon new set of conditions. In all cases of wind system planning, it should be clear that you must investigate restrictions on your land title, zoning restrictions in your area, and requirements your building codes may place on the project. A good way to start that investigation is to apply for a building permit for the project. That should help bring all the little goblins out of the closet.

Once you've investigated all of your rights by assessing the various restrictions, you might ask yourself one last question: "Do I have the right to the wind energy here?" You answer this by determining if it is possible that your neighbor, or his or her successor, might build a high-rise condominium just upwind of your newly installed wind machine. If you think it is possible, try to get a deeded restriction placed on that property to preclude such a possibility. When you try this gambit, you soon get involved in the great issue of wind and solar rights. Old English law said that one had the right to the sunshine that fell upon his property. Old English law failed to notice that a neighbor's high-rise might block that sun so that no, or less, sun would fall upon the property in question.

New American law is slowly taking up questions of wind and solar rights. Several states have spelled out methods for resolving competing claims to the sun, but, as of mid-1982, only Oregon had incorporated wind rights into state law. Oregon's 1981 law gives local landowners the right to control the wind passing over their property. In this way it establishes a means for local government to treat wind in its planning. It will be awhile before other states adopt similar legislation; you're not alone if you discover you might have a problem.

Although you may have the right to install a wind machine, you gain obligations in the course of that project. These obligations focus on your protecting the health and well being of the general populace. Now, you might not feel much like protecting somebody's well being when you arrive home from a shopping trip to discover absolute strangers climbing your tower, but you must. This obligation translates directly to liabilities. Just what you might be liable for in the event of a problem is certainly a good question to ask your attorney, but in any case it's best to plan a course of careful preventive action.

Preventive action starts with a sound installation. It continues with religious attention to maintenance and inspections (you don't drive the freeways on bald tires, do you?), and ends by keeping the public away from your wind machine, except when invited by you. This calls for a tall safety fence much like you would put around a swimming pool. Be sure to include the expense of this fence in your cost estimates.

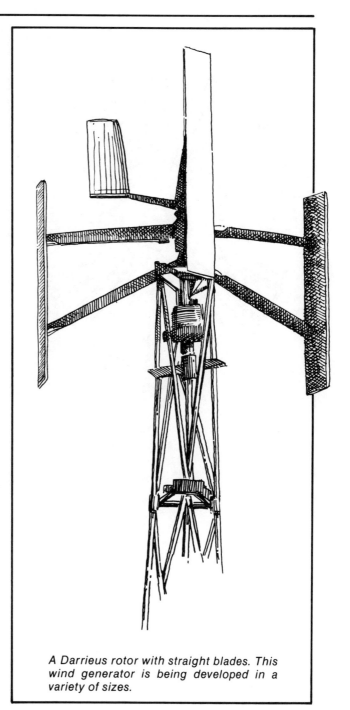

A Darrieus rotor with straight blades. This wind generator is being developed in a variety of sizes.

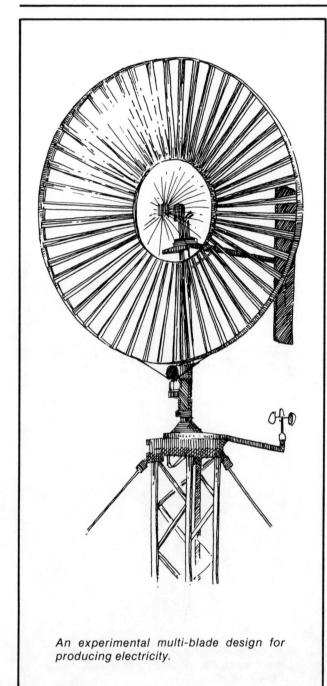

An experimental multi-blade design for producing electricity.

Frequently, in order to qualify for tax credits, the wind machine you purchase must meet certain requirements. For example, it may have to offer a three-year warranty. Such a warranty is worthless if the manufacturer or installer is not around to provide the service when you need it, which adds another dimension to carefully selecting your equipment. Along with a warranty, you might consider a maintenance contract; this is perhaps the best way to ensure your wind system gets serviced regularly.

SOCIAL ISSUES

The primary social issues you must face may already be folded up neatly in a set of architectural restrictions on your property, or in your local zoning laws. They may not entirely, however, be reflected in those restrictions. Your neighbors, with whom you ordinarily play bridge, may have some powerful feelings about you or anybody owning a wind machine on the block. These feelings are usually based on preconceived notions about problems associated with wind machines. Noise, television interference, and general aesthetics, not to mention overwhelming concern for the village's little folks, are the usual concerns. They may be founded in local experience, or they may come from badly researched newspaper stories. In any event, the best way to maintain a happy bridge club is to investigate your neighbors' feelings before you install your wind machine.

You may want to go further. Your neighbors may be willing to share the costs and benefits of a wind machine. It may make good economic sense to cooperate and hook up two or more homes to one machine or even to establish a neighborhood wind farm.

Many windmills are not much noisier than the wind that powers them. Some are quite noisy, however, especially if they run out of transmission oil. It will pay to assess the noise situation when talking with your dealer and installer.

Television interference is another issue. Medium and large wind machines with metal blades may scatter TV signals and fuzz up local TV sets. Carefully placed, a small machine may, at worst, only affect the owner's TV reception. Your dealer and installer will have the latest information on the TV interference of the wind machines you may choose.

Aesthetics is a whole story in itself, best left to individuals to decide. The authors happen to think all wind machines are creative, dynamic art forms. Surely, your neighbors can develop a healthy taste for kinetic art!

When you have finally investigated all of these issues, when you are satisfied that it is possible to install a wind machine and live with it, and when you have the budget available, the hard part begins. You have to jump off the bridge and have the machine installed. Chapter 6 discusses some appropriate bridge-leaping techniques.

GUIDE 5

ECONOMIC EVALUATION OF A WIND INSTALLATION

After taking wind speed measurements for a full year, a farmer determined that his site had a 14-mph annual average. Using the information in Guide 4, he estimated that the 45-feet diameter wind machine he was interested in could produce 73,000 kWh annually. The dealer quoted him $38,000 for the wind machine, tower, and control system. Estimates for installation and a warranty contract came to $7,000. The total installed cost was an estimated $45,000. The farmer's figure of merit for this installation was (73,000 kWh/$45,000) = 1.62 kWh/$.

The farmer opted to sell all the electricity generated by the windmill back to the utility company at a rate of 6.5 cents per kWh. With this, he expected to generate additional revenue of $4,745 in the first year. He also expected that the utility company's avoided cost would continue to increase at a slightly more rapid rate than general inflation so that in future years the wind machine would generate even more revenue.

Next, the farmer talked to his accountant and banker and determined that, with the right terms, he could make the payments on a loan for $25,000 out of the revenue generated by the windmill. By doing this, the loan payments were treated as one more business expense which could be paid by the revenues from sales of electricity. Money left over after expenses would be additional personal income. The remaining $20,000 was the farmer's own equity investment.

Life cycle costing methods are usually used to compare several options that are open to a person spending his money. In the following example, the present value of the farmer's $20,000 investment in the windmill is compared to the same investment in bonds.

GUIDE 5

LIFE CYCLE COST COMPARISON OF WINDMILL INVESTMENT WITH BONDS

Investment amount: $20,000
Options: windmill or bonds

	Present Value Factor (2)	Windmill Income $	Windmill PV (3) $	Bonds Income $	Bonds PV (3) $
After Tax Return:					
Year 1	1.150	23,315	20,274	1,680	1,461
Year 2	1.322	(265)	(200)	1,680	1,281
Year 3	1.521	3,157	2,076	1,680	1,105
Year 4	1.749	3,376	1,930	1,680	961
Year 5	2.011	3,756	1,868	1,680	835
Total Return		33,339	25,948	8,400	5,643
Internal Rate of Return		36.2%			
Annual Yield (1)				8.4%	

Notes: (1) Return on bonds is 14%, which yields 8.4% after federal and state taxes. After-tax return on windmill taken from cash-flow analysis.

(2) Present value factor is based on 15% discount rate, and is calculated as $(1.15)^t$, where t is the year.

(3) Present value of after-tax return is calculated as (Income $)/(Present Value Factor).

After five years, the windmill has produced an after-tax income with a present value of $33,339, while the bond investment of the same amount produced $8,400. The annual windmill income drops off significantly in future years due to loss of the federal depreciation tax benefits. Nonetheless, over a 10- to 20-year investment period, the windmill will still have a higher cash return as well as a higher rate of return.

Another way of accounting for costs and benefits that is frequently used for income-producing projects is called cash-flow analysis. The following cash flow for this project details the method of accounting for costs over the first five years of the project. The analysis shows that the owner invested $20,000 at the start of the project and recovered $23,315 as income, after taxes, by the end of the first year.

The enclosed cash-flow statement is an example that applies to one specific project. Tax laws and other conditions change with time and location. Expert advice is recommended for this type of investment. Several references for further reading on this subject are listed in the Bibliography.

GUIDE 5

For a 45° Diameter Wind Machine at a 14-mph Location

Installed Cost: $45,000 Loan Amount: $25,000 Equity Amount: $20,000
Energy Production: 73,000 kWh/yr   Energy Sales at 6.5 c/kWh in Year 1

CASH-FLOW ANALYSIS

		Year 1	Year 2	Year 3	Year 4	Year 5
A.	Income	4,745	5,220	5,741	6,316	6,947
B.	Expenses:					
	D&M	203	219	236	255	275
	Prop. taxes	551	562	573	585	596
	Insurance	122	131	142	153	165
	Total expenses	876	912	951	993	1,036
C.	Debt service	4,275	4,275	4,275	4,275	4,275
D.	Cash flow before taxes	(406)	33	515	1,048	1,636
E.	Federal tax credits	9,563	0	0	0	0
F.	State tax credits	10,125	0	0	0	0
G.	Federal depreciation (ACRS)	6,750	9,900	9,450	9,450	9,450
H.	State depreciation (DB)	23,250	7,750	2,583	0	0
I.	Interest part of debt ser.	3,750	3,671	3,580	3,476	3,356
J.	Federal taxable income or (Loss) A-B+F-G-I-M	(6,631)	2,712	(7,673)	(7,493)	(7,043)
K.	State taxable income or (Loss) A-B-H-I	(23,131)	(7,083)	(1,373)	1,847	2,556
L.	Federal income tax liability or (Refund)	(2,183)	895	(2,532)	(2,473)	(2,324)
M.	State income tax liability or (Refund)	(1,850)	(567)	(110)	148	204
N.	Net cash flow after taxes D+E+F-L-M	23,315	(265)	3,157	3,373	3,756

Internal rate of return on net cash flow after taxes (5 years) = 36.2%

ASSUMPTIONS:

Tax losses are assured to offset other tax liability of owner. Owner assumed to be in a 33% federal tax bracket and 8% state tax bracket.

Income: Based on 73,000 kWh/yr, $.065/kWh in year 1, increasing at 10%/yr.

Expenses: Escalation rates are: 8%/yr on D&M, 2%/yr on property tax, 8%/yr on insurance.

Debt Service: $25,000 loan at 15% for 15 years.

GUIDE 5

Tax Credits: Federal tax credits include a 10% Investment Tax Credit (ITC) and 15% Energy Property Investment Tax Credit. State energy tax credits of 25% are applicable. These credits apply toward most expenses in a project. For this case, 85% for federal and 90% for state was assured. State tax credits are treated as income for federal tax purposes.

Equipment Depreciation: The 1981 Accelerated Cost Recovery System is applied in the example. Wind equipment qualifies for 5-yr. depreciation. It is assured that 85% of project costs form the basis for depreciation. State depreciation applies toward that part of investment not claimed as a tax credit. Three-year double-declining balance method used.

Federal Income Tax: Calculated as income (A) less expenses (B) plus state tax credit from previous year (F) less federal depreciation (G) less interest portion of debt service (I) less state income tax liability from previous year (M). Tax rates vary, up to 50% depending on income and deductions.

State Income Tax: Tax rates vary, depending on income and deductions. In California they can be as high as 11%; 8% used for this example.

Chapter 6

BUYING, INSTALLING, AND OWNING A WIND SYSTEM

Wind site analysis, load estimation, windmill selection, warranties, liabilities, noise—a host of items to deal with along the path to planning, installing, and owning a wind system. These factors really aren't all that much when you think about it. After all, you go through much the same mental trauma when you purchase a new car; the only difference is that new car dealers are—or used to be—planted on just about every main intersection. Wind energy dealers are not.

Because (as this is written) a vast infrastructure of manufacturers, distributors, dealers, finance companies, insurance companies, and network advertising is not in place, the whole process seems complicated, if not downright arcane. "Why not just go out and buy a wind system, install it, and see what it will do?" Well, why not? There is probably no good reason you should not do just that, especially if you purchase a small system and you promise not to be disappointed.

If, however, you have already ground up a few pencils figuring out what wind energy will cost, chances are you're apt to land somewhere between pleasantly surprised and thoroughly ashamed of yourself if you ignore any of the steps involved in wind system planning and jump off the bridge without measuring the depth of the water.

In the early days of wind energy, if it was windy—and there was never a question if it was—a windcharger was often installed. As the windy areas became sold out, dealers ventured into less windy areas. They did not, in those days, pay much attention to anemometers, or data collection. Recent interest in wind energy has been at once encouraged and dampened by a desire to collect data. You must expect to perform some data collection or get someone to perform it on your behalf.

Recent interest in wind energy has brought about a return of wind manufacturers and dealers. Thus, the infrastructure is in the making. For example, California has—at this writing—at least six and maybe more bona fide windmill manufacturers, which makes the state the most active in this field. So help is available in planning your wind system.

The final act in the process is signing a purchase order for a windmill to be installed on your property. State-licensed contractors will be required if you hire out the installation; a state-licensed engineer may also need to review the installation plans for some county building code requirements. Your purchase specifications should deal not only with the wind machine, but also with the warranty, completion bonds on the installation, lien waivers for all workers, and perhaps future maintenance agreements. These are all normal considerations if you buy a swimming pool; a windmill installation should not be treated any differently. If you deal with a wind machine intelligently and thoroughly, there's no reason why you won't be satisfied.

TYPICAL INSTALLATION

"Just what is an installation procedure like? What sort of maintenance is required?" These two facets of ownership are the two most often questioned aspects of wind power after "Will it power my house?" There is probably no "typical" installation, and what follows may in no way describe your situation.

A site has been chosen where the foundation will be planted to support the tower. A large hole—or several smaller ones—are dug. A building inspector checks to see that the hole is equal to the approved drawings. It rains; the hole collapses and must be redug—but, you hope not re-inspected. It's about to rain again.

Concrete, rebar, foundation bolts, parts of the tower, or the entire tower are all planted and secured to allow the concrete to cure for a few days. Then the wind machine is hoisted aloft.

There are several ways to hoist a windmill, none of them particularly easy. One way is to winch the machine up the tower as a complete assembly—a cherry picker or crane is often substituted for a block and tackle. Another way is to winch the machine up the tower as a series of subassemblies, to be assembled and adjusted aloft. This is much like overhauling your car engine eighty feet above the ground, but it works and is good exercise.

A third way is to assemble the machine on the ground and install it on the tower, which is lying down. The tower is then tilted up, wind machine and all. This method is cleanest but requires the most land for the tilting operation, and if it is done improperly, the tower might collapse about half-way through the tilting process. Some windmill manufacturers are designing their products specifically to be installed this way. The primary advantage is that the machine can be tilted back down for maintenance (or in anticipation of gale winds).

Once the wind machine is installed, you must fight off an attack of "fire-em-up-itis," an infectious disease that usually starts with ". . . wouldn't it be fun to spin 'er up a while?" and ends with ". . . who was responsible for that missing bolt?" and ". . . wow, it sure hit the ground hard!" The entire installation must be completed and a careful test and adjustment plan implemented; the first spin-up should be run in very gentle breeze.

TYPICAL MAINTENANCE

Wind machine maintenance is just about the same as car maintenance: some items need to be oiled or greased, others need to be inspected periodically, and perhaps a few items should be replaced at specified times. There are no tires to wear out, no road grease to gunk up, but dirt will accumulate in your wind machine. Dirt contributes to bearing, slip-ring, and brush wear.

Manufacturers specify the required maintenance for their windmills. This usually includes a seasonal oil change, greasing the bearings, annually cleaning all the innards, and careful inspection at least twice a year. That is for the equipment on top of the tower. The tower itself should be checked at least annually for bolt tightness and evidence of damage.

Other equipment may never need checking, especially the electronic boxes. If you do open them up, watch out for high voltage.

Batteries, on the other hand, are like water-well piston pumps: they need occasional inspection, and eventually they just wear out. Batteries need to be protected from shorts (tools falling across the terminals). They need to be cleaned outside and their fluid level checked occasionally, more often if cycled heavily.

These maintenance items are actually less complex than the ones you perform on your auto. Even so, you may wish to leave them to experienced technicians, just like most people do with their cars. A maintenance contract is usually available from a dealer for such purposes. Climbing a tower to view the neighbor's swimming pool might be amusing but maintenance eighty feet up requires skill and experience. Where possible, maintenance is best left to qualified technicians. Where not, don't look down.

CONCLUSIONS

If you have read this entire book, you are now familiar with the various aspects of planning and owning a wind system. Perhaps you didn't expect the process to seem quite so complex. Perhaps you thought it would be worse. In either case, you are now armed with enough information to deal intelligently with your friendly local wind system dealer.

But don't expect buying a wind machine to be the same as buying a car. You cannot drop by a showroom and pick one in lavender with a yellow racing stripe and long blades. Wind machines are not now, and likely never will be, sold that way. If you want special colors, plan to paint it yourself; many people do.

Where to go from here? Two approaches are sensible at this point in your project: buy a machine, or gather more information. Most people tend toward the latter approach (some, in fact, never leave the information-gathering mode). The authors tend toward the former, which is why they are authors, not book collectors. More information is indeed necessary, however, since this book was designed as a common-sense introduction to planning a wind system. Some other useful books are listed in the Bibliography. These books may be available at or through your local bookstore; they should also be available at your local solar and wind systems dealer.

In addition to reading books, it's a good idea to visit a few wind installations in your area. This is a fine way to become familiar with your windmill dealer and his or her performance with other clients. It's also a good way to see if there's much wind energy available in your area. Naturally, if nobody else has jumped off the bridge, you get to be first.

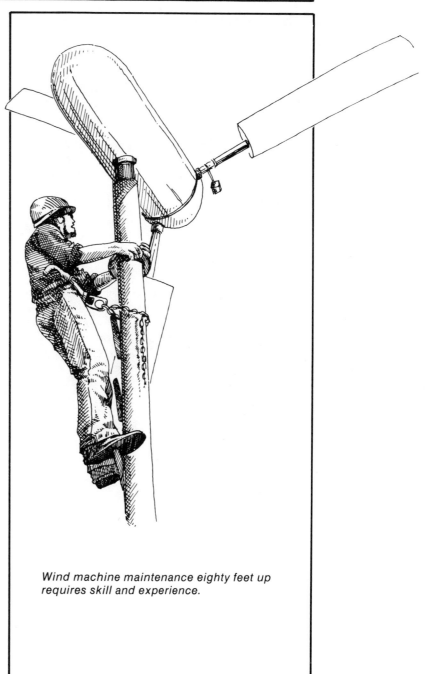

Wind machine maintenance eighty feet up requires skill and experience.

CASE STUDIES

CASE STUDY 1
Merle Tate, Common Falls, Minnesota

I have two wind machines on my farm. One is an old Jacobs that I first put up in the early 1940s and the other a new Jacobs I put up four years ago. Both are running, doing different jobs, and doing them well.

The old machine taught me about the potential of wind. I ran most of my equipment with it: I pumped water, ran a vacuum pump for my milking machine, powered a refrigerator, toaster, washing machine, and a radio. I couldn't, however, run two half horsepowers at the same time, so I had to zig-zag a bit so as not to overload the batteries. I put that machine away for a long time, 30 years or more, and then decided I could use it for auxiliary heat. I set up the generating tower again, then I wired it to a portable heater. It's a resistance heater, 220 volts—3000 watts, that cost me about $35. To make it work, I had to split the heating element into five parts. That made about 40 volts for each split. I hooked them up in parallel, and that made about 40 volts. So that heater today is taking all the generator will put out. It has been providing auxiliary heat for one room for the past five or six years.

Curious about how much energy the machine was generating, I put a direct-burn per kilowatt meter on the generator. I found out it was putting out about 300-kilowatt hours a month. That showed me the merit, so I went down to Florida to see Mr. Jacobs about putting up another machine. I came back and used the statistics from the local weather bureau that showed an average wind speed through here of about 12 mph. I had no trouble putting the machine up, I got help from Red Wing Voc Tech, but it took me about a year and a half to convince the Rural Energy Association, a cooperative, to buy my surplus electricity. They weren't too happy at first, but by patience I guess, I got the agreement, a year and a half before there was any such thing as PURPA.

First, they told me I didn't have any background, that I didn't know enough about wind energy. They were also afraid of energizing the line; if the line was down they felt it could injure somebody working on it. We proved to them that it couldn't energize any current unless the line was alive, because the line energizes the generator. In other words, if the line is not in operating condition and it goes off, then my machine can't generate any electricity at all.

I still had to meet with the co-op board three or four times. They suggested I go to their wholesaler, the Cooperative Power Association. I also knew Governor Quie a little personally, and he helped me get all the way to the administrator of the Rural Energy Association in Washington. But I got a letter from them, and they didn't think too much of it. And even the governor then thought it should take five or six years until the field developed enough for the power company to take my electricity. They were also afraid of the liability as far as the safety of the line.

Finally though, my approach was this: I told them that it takes a pound of coal to deliver one kilowatt hour of power. Well, if my old machine was producing 300 kilowatts a month, that is 300 pounds of coal that doesn't have to be pumped out of the ground. And the new machine, that is producing 1,000 to 1,800 kilowatts a month, is *really saving* in coal, fossil fuel, or atomic power. So we finally came to terms.

I just made a tentative arrangement with them. They wanted me to sign a five-year agreement, but I told them I wouldn't because there would be too many changes. They are allowing me the wholesale cost and I am putting all of it in the line at wholesale cost. That way I am supporting my co-op by using their line, and I am not taking advantage of my neighbor who doesn't own a wind machine.

Wind is not the total answer as far as total electric power is concerned. But it could supply half of the consumption in Goodhue County where I live. We have about 3,500 customers and the average consumption is about 1,500 kilowatt hours a month, including residences, farms and some commercial uses. My plant is making at least half of the average consumer's consumption. That may be unbelievable, but that's what's happening.

CASE STUDY 2
John Wolfe, Barnstable, Massachusetts

The combination of a wind machine and a wood stove has decreased electrical use in my all-electric home by 10,000 kWh since the mid-1970s.

I always knew there was enough wind on the north side of Cape Cod to generate electricity. My living room anenometer has measured 86-mph winds coming off the bay. In 1976, when I read an ad in the local newspaper about an Enertech wind machine that could be plugged into an electrical system without batteries, I decided that was for me.

It took about five months to get permission from the local historic district commission to construct the wind machine; I went to each abutter and explained what I was going to do, and they were all for it. The machine itself was very easy to put up: we dug a 10-foot hole for the 60-foot tower and, with a crane, had the 275-pound machine mounted on top.

Most of our wind comes from the north; there is very little southeast or southwest wind, and it gets dissipated by the trees and turbulence. Summertime wind is also very poor. In one July, in fact, I only made ten kWh.

The best month I've ever had was December of 1981 when I had 401 kWh and the previous January I had 391 kWh. In two and a half years, I have had six months with 200 kWh or better. My electrical consumption has decreased from 16,000 kWh in 1974 to 6,400 this year, which I attribute mainly to the wind machine and a wood stove I installed last year.

The electric company has been cooperative even though they did not have a buy-back policy when I started. There has been a lot of federal red tape and papers to sign that, quite frankly, I didn't always understand.

When I first started, I ran my meter backward when the wind machine was running, then they put in a meter that wouldn't go backward. After that, they tried a double-headed meter that measured what went into my house and what came out. Now the meter runs both forward and backward. I am billed for the difference at the retail rate.

Right now the utility is not losing much, and I am not making much. My machine is only designed for two kilowatts. I figure you need a ten kilowatt machine in this part of the country to break even with the utility company. Also, I had some trouble with the machine—some bolts in the nose came loose last year. Fortunately, the manufacturer replaced the whole machine for me.

But those are minor problems. Would I suggest someone else follow my example and put up a small wind machine? Yes. Eventually it is going to come to this for everybody—maybe not wind, but some form of solar energy. I don't believe we have an inexhaustible oil supply.

APPENDIX 1

POWER AND ENERGY REQUIREMENTS OF APPLIANCES AND FARM EQUIPMENT

Name	Watts	Hrs/Mo	KWHRS/Mo
Air conditioner, central			620*
Air conditioner, window	1566	74	116*
Battery charger			1*
Blanket	190	80	15
Blanket	50-200		15
Blender	350	3	1
Bottle sterilizer	500		15
Bottle warmer	500	6	3
Broiler	1436	6	8.5
Clock	1-10		1-4*
Clothes drier	4600	20	92*†
Clothes drier, electric heat	4856	18	86*†
Clothes drier, gas heat	325	18	6*†
Clothes washer			8.5*
Clothes washer, automatic	250	12	3*
Clothes washer, conventional	200	12	2*†
Clothes washer, automatic	512	17.3	9*
Clothes washer, ringer	275	15	4*†
Clippers	40-60		½
Coffee maker	800	15	12
Coffee maker, twice a day			8
Coffee percolator	300-600		3-10
Coffee pot	894	10	9
Cooling, attic fan	1/6-3/4 HP		60-90*†
Cooling, refrigeration	3/4-1½ ton		200-500*
Corn popper	460-650		1
Curling iron	10-20		½
Dehumidifier	300-500		50*
Dishwasher	1200	30	36*
Dishwasher	1200	25	30*
Disposal	375	2	1*
Disposal	445	6	3*
Drill, electric, ¼"	250	2	5
Electric baseboard heat	10,000	160	1600
Electrocuter, insect	5-250		1*
Electronic oven	3000-7000		100*
Fan, attic	370	65	24*†
Fan, kitchen	250	30	8*†
Fan, 8"-16"	35-210		4-10*†
Food blender	200-300		½
Food warming tray	350	20	7
Footwarmer	50-100		1
Floor polisher	200-400		1
Freezer, food, 5-30 cu ft.	300-800		30-125*
Freezer, ice cream	50-300		½
Freezer	350	90	32*
Freezer, 15 cu. ft.	440	330	145*
Freezer, 14 cu. ft.			140*
Freezer, frost free	440	180	57*
Fryer, cooker	1000-1500		5
Fryer, deep fat	1500	4	6
Frying pan	119	12	15
Furnace, electric control	10-30		10*
Furnace, oil burner	100-300		25-40*
Furnace, blower	500-700		25-100*†
Furnace, stoker	250-600		3-60*†
Furnace, fan			32*†
Garbage disposal equipment			
	1/4-1/3 HP		½*
Griddle	450-1000		5
Grill	650-1300		5
Hair drier	200-1200		½-6*
Hair drier	400	5	2*
Heat lamp	125-250		2
Heater, aux.	1320	30	40
Heater, portable	660-2000		15-30
Heating pad	25-150		1
Heating pad	65	10	1
Heat lamp	250	10	3
Hi Fi Stereo			9*
Hot plate	500-1650		7-30
House heating	8000-15,000		1000-2500
Humidifier	500		5-15*
Iron	1100	12	13
Iron			12
Iron, 16 hrs/month			13
Ironer	1500	12	18
Knife sharpener	125		¼*
Lawnmower	1000	8	8*†
Lighting	5-300		10-40
Lights, 6 room house in winter			60

67

APPENDIX 1

Name	Capacity (watts)		Est. KWHR
Light bulb, 75	75	120	9
Light bulb, 40	40	120	4.8
Mixer	125	6	1
Mixer, food	50-200		1
Movie projector	300-1000		
Oil burner	500	100	50*
Oil burner			50*
Oil burner, 1/8 HP	250	64	16*
Pasteurizer, ½ gal.	1500		10-40
Polisher	350	6	2
Post light, dusk to dawn			35
Power tools			3
Projector	500	4	2*
Pump, water	450	44	20*†
Pump, well			20*†
Radio			8
Radio, console	100-300		5-15*
Radio, table	40-100		5-10*
Range	8500-1600		100-150
Range, 4 person family			100
Record player	75-100		1-5
Record player, transistor	60	50	3*
Record player, tube	150	50	7.5*
Recorder, tape	100	10	1*
Refrigerator	200-300		25-30*
Refrigerator, conventional			83*
Refrigerator-freezer	200	150	30*
Refrigerator-freezer, 14 cu.ft.	326	290	95*
Refrigerator-freezer, frost free	360	500	180*
Roaster	1320	30	40
Rotisserie	1400	30	42*
Sauce pan	300-1400		2-10
Sewing machine	30-100		½-2
Sewing machine	100	10	1
Shaver	12		1/10
Skillet	1000-1350		5-20
Skil Saw	1000	6	6
Sunlamp	400	10	4
Sunlamp	279	5.4	1.5
Television	200-315		15-30*
TV, BW	200	120	24*
TV, BW	237	110	25*
TV, color	350	120	42*
TV, color			100*
Toaster	1150	4	5
Typewriter	30	15	0.5*
Vacuum cleaner	600	10	6
Vacuum cleaner, 1 hr/wk			4
Vaporizer	200-500		2-5
Waffle iron	550-1300		1-2
Washing machine, 12 hrs/mo			9*
Washer, automatic	300-700		3-8*
Washer, conventional	100-400		2-4*
Water heater	4474	89	400
Water heater	1200-7000		200-300
Water pump (shallow)	½ HP		5-20*†
Water pump (deep)	1/3-1 HP		10-60*†

AT THE BARN

Name	Capacity HP or watts	Est. KWHR
Barn cleaner	2-5	120/yr*
Clipping	fractional	1/10 per hr.
Corn, ear crushing	1-5	5 per ton*
Corn, ear shelling	¼-2	1 per ton*†
Electric fence	7-10 watts	7 per mo.*†
Ensilage blowing	3-5	½ per ton
Feed grinding	1-7½	½-1½ per 100 lbs.*†
Feed mixing	½-1	1 per ton*†
Grain cleaning	¼-½	1 per ton bu*†
Grain drying	1-7½	5-7 per ton*†
Grain elevating	¼-5	4 per 1000 bu*†
Hay curing	3-7½	60 per ton*
Hay hoisting	½-1	1/3 per ton*†
Milking, portable	¼-½	1½ per cow/mo.*†
Milking, pipeline	½-3	2½ per cow/mo.*†
Sheep shearing	fractional	1½ per 100 sheep
Silo enloader	2-5	4-8 per ton*
Silage conveyer	1-3	1-4 per ton*
Stock tank heater	200-1500 watts	varies widely
Yard lights	100-500 watts	10 per mo.
Ventilation	1/6-1/3	2-6 per day*† per 20 cows

IN THE MILKHOUSE

Name	Capacity HP or watts	Est. KWHR
Milk cooling	½-5	1 per 100 lbs. milk*
Space heater	1000-3000	800 per year
Ventilating fan	fractional	10-25 per mo.*†
Water heater	1000-5000	1 per 4 gal

FOR POULTRY

Name	Capacity HP or watts	Est. KWHR
Automatic feeder	¼-½	10-30 KWHR/mo*†
Brooder	200-1000 watts	½-1½ per chick per season
Burglar alarm	10-60 watts	2 per mo.*
Debeaker	200-500 watts	1 per 3 hrs.

APPENDIX 1

Egg cleaning or washing	fractional	1 per 2000 eggs*†	
Egg cooling	1/6-1	1¼ per case*	
Night lighting	40-60 watts	10 per mo. per 100 birds	
Ventilating fan	50-300 watts	1-1½ per day*† per 1000 birds	
Water warming	50-700 watts	varies widely	
FOR HOGS			
Brooding	100-300 watts	35 per brooding period/litter	
Ventilation fan	50-300 watts	¼-1½ per day*†	
Water warming	50-1000 watts	30 per brooding period/litter	
FARM SHOP			
Air compressor	¼-½	1 per 3 hr.*	
Arc welding	37½ amp	100 per year*	
Battery charging	600-750 watts	2 per battery charge*	
Concrete mixing	¼-2	1 per cu. yd.*†	
Drill press	1/6-1	½ per hr.*†	
Fan, 10"	35-55 watts	1 per 20 hr.*†	
Grinding, energy wheel	1/4-1/3	1 per 3 hr.*†	
Heater, portable	1000-3000 watts	10 per mo.	
Heater, engine	100-300 watts	1 per 5 hr.	
Lighting	50-250 watts	4 per mo.	
Lathe, metal	¼-1	1 per 3 hr.	
Lathe, wood	¼-1	1 per 3 hr.	
Sawing, circular 8"-10"	1/3-1/2	1/2 per hr.	
Sawing, jig	1/4-1/3	1 per 3 hr.	
Soldering, iron	60-500 watts	1 per 5 hr.	
MISCELLANEOUS			
Farm chore motors	½-5	1 per HP per hr.	
Insect trap	25-40 watt	1/3 per night	
Irrigating	1 HP up	1 per HP per hr.	
Snow melting, sidewalk and steps, heating-cable imbedded in concrete	25 watts per sq. ft.	2.5 per 100 sq. ft. per hr.	
Soil heating, hotbed	400 watts	1 per day per season	
Wood sawing	1-5	2 per cord	

Symbol Explanation
*AC power required
†Normally AC, but convertible to DC

Notes: Lighting in this table is assumed to be incandescent — if fluorescent, the wattage bulbs consume the same power but deliver 3 times as much light — fluorescent bulbs also require AC, but can be converted to DC.

These figures can be cut by 50% with conservation of electricity.

1) Sources for this table represent a conglomerate of several separate tables taken from:
   a) Northern States Power Co., Mpls. Mn
   b) University of Minnesota, Agricultural Extension Service
   c) Seattle City Light, Seattle, Washington
   d) Energy Conservation Techniques, National Bureau of Standards
   e) Garden Way Labs
   f) Henry Clews
   g) Real Gas and Electric

"Composite of Kilowatt Hour Ratings for Various Appliances" excerpted from **Energy Primer: Solar, Water, Wind and Biofuels**. Updated and revised edition, edited by Richard Merrill and Thomas Gage. Copyright (c) 1978 by Portola Institute. Reprinted by permission of Dell Publishing Co., Inc.

APPENDIX 2

ELECTRICAL WIND MACHINE MANUFACTURERS
Compiled by Vicki Butler,
Office of Appropriate Technology, California Energy Commission

Aerolite
550 Russells Mills Road
P.O. Box 576
South Dartmouth, MA 02748
(617) 638-8722
6 KW, 10 KW, 14 KW

Aero Power Systems, Inc.
2398 - 4th Street
Berkeley, CA 94710
(415) 848-2710
1 KW, 1.5 KW

Aerotherm Corporation
P.O. Box 574-A, Route 1
Lenhartsville, PA 19534
(215) 562-5622
25 KW

Altos—The Alternate Current
c/o Triton Tool & Manufacturing
5420 Arapahoe Road
Boulder, CO 80303
(303) 442-0885
1.5 KW

American Energy Savers
912 St. Paul Road, Box 1421
Grand Island, NB 68801
(308) 382-1831
8 KW

Astral Wilcon
127 W. Main Street
Millbury, MA 01527
(617) 865-9570
10 KW

Bendix Corporation
Energy, Environment &
Technology Office
2582 S. Tejon
Englewood, CO 80110
(303) 922-6394
4.5 KW

Bergey Wind Power Co.
2001 Priestly Avenue
Norman, OK 73069
(405) 364-4212
1 KW

Bertoia Studio
644 Main Street
Bally, PA 19503
(215) 845-7096
4 KW

Bircher Machine, Inc.
Box 97
Kanopolis, KS 67454
(913) 472-4413
5 KW

**Boeing Engineering &
Construction**
Wind Energy &
Environmental
P.O. Box 3707
Seattle, WA 98124
(206) 575-5985
2500 KW

Carter Wind Systems, Inc.
P.O. Box 405-A
Burkburnett, TX 76354
(817) 569-2238
25 KW

Chalk Wind Systems
P.O. Box 446
St. Cloud, FL 32769
(305) 892-7338
1 KW, 1.5 KW

Coulson Wind Electric, Inc.
RFD 2, Box 225
Polk City, IA 50226
(515) 984-6038
15 KW

Dragonfly Wind Electric
P.O. Box 57
Albion, CA 95410
(707) 937-4710
300 W

Elfin Corporation
550 Chippenhook Road
Wallingford, VT 05773
(802) 446-2575
10 KW (Prototype)

Engineering Model Lab
8 Waterhouse Road
P.O. Box 293
Buzzards Bay
Bourne, MA 02532
(617) 759-7659
5 KW (Prototype)

APPENDIX 2

Environmental Energies, Inc.
Front Street
Copemish, MI 49625
(616) 378-2921
6 KW, 9 KW, 12 KW, 15 KW
(All are Prototypes)

ESI (Energy Sciences, Inc.)
900 - 28th Street
Boulder, CO 80303
(303) 449-3559
50 KW

Fayette Manufacturing Corp.
712 River Street
Clearfield, PA 16830
(814) 765-1631
20 KW, 50 KW

Fel-Pro Energy, Inc.
P.O. Box 27
Lake Geneva, WI 53147
(414) 248-6672
4 KW (Prototype)

FloWind Corp.
21414 - 68th Avenue South
Kent, WA 98031
(206) 872-8500
100 KW (Prototype)

General Electric Co.
P.O. Box 527
King of Prussia, PA 19406
(215) 962-1219
6200 KW (Prototype)

Grumman Allied Industries, Inc.
445 Broad Hollow Road
Melville, NY 11747
(516) 454-8678
15 KW (Prototype)

Hamilton Standard
Windsor Locks, CT 06906
(203) 623-2800
3 MW, 4 MW (Prototypes)

Jacobs Wind Electric Co.
2720 Fernbrook Lane
Minneapolis, MN 55441
(612) 559-9361
10 KW

Jerico
915 - 8th Street, S.E.
Detroit Lake, MN 56501
(218) 847-6631
6 KW

Lockheed, Palo Alto Research
3251 Hanover Street
Palo Alto, CA 04304
(415) 493-4411, ext. 45007
(Currently not manufacturing but willing to share information about their discoveries)

Megatech Corp.
29 Cook Street
Billerica, MA 01866
(617) 273-1900
200 W (One part of an educational package)

Millville Hawaii Windmills, Inc.
3028 Ualena Street
Honolulu, HI 96819
(808) 836-1180
15 KW

North Wind Power Co.
P.O. Box 556
Moretown, VT 05660
(802) 496-2955
2 KW (6 KW Prototype)

Oakridge Windpower, Inc.
P.O. Box 634
107 East Henning Street
Battlelake, MN 56515
(218) 864-8403
10 KW, 20 KW

Pawl Inventioneering
1325 Delta Road
Walled Lake, MI 48188
(313) 624-5597
250 W, 1 KW, 4 KW

PM Wind Power Inc.
P.O. Box 89
Mentor, OH 44060
(216) 255-3437
25 KW

Product Development Institute
445 Talmadge Road
Toledo, OH 43623
(419) 472-2136
6.5 KW

Pinson Energy Corporation
P.O. Box 7
Marstons Mills, MA 02648
(617) 428-8535
5 KW

Power Group International Corp.
13315 Steubner-Airline Road
Suite 106
Houston, TX 77014
6 KW (Prototype)

Sancken Wind Electric, Inc.
4140 Skylark Road
Kingman, AZ 86401
(602) 757-2526
1 KW, 5 KW, 10 KW

Sencenbaugh Wind Electric
253 Polaris
Mt. View, CA 94043
(415) 964-1593
500 W, 1 KW, 2 KW

S. J. Windpower
c/o Nielson Iron Works
P.O. Box 128
Racine, WI 53401
(414) 632-7571
10 KW

Soma Windmills (New Zealand)
c/o Appropriate Power Systems
1095 Meadowvale Road
Santa Ynez, CA 93460
(805) 688-8590
600 W, 1 KW

U.S. Windpower, Inc.
160 Wheeler Road
Burlington, MA 01803
(617) 273-4502
50 KW (Machine not for sale; they sell electricity)

APPENDIX 2

Wecs-Tech Corp.
4327 Redondo Beach Boulevard
Lawndale, CA 90260
(213) 542-1666
70 KW (100 KW Prototype)

Wenco (Switzerland)
c/o Windtex, Inc.
200 Rupert Street
Fort Worth, TX 76107
(817) 332-6352
60 KW, 100 KW

Westinghouse Electric Corp.
875 Greentree Road
Bldg. 8, M.S. 414
Pittsburgh, PA 15220
(412) 928-2430
500 KW (Prototype)

Whirlwind Power Co.
5030 York Street
Denver, CO 80216
(303) 595-8491
3 KW

Winco
Division of Dyna Technology
7850 Metro Parkway
Minneapolis, MN 55420
(612) 853-8400
200 W, 450 W

Windcraft Energy Systems, Inc.
P.O. Box 131
Goessel, KS 67053
(316) 367-8222
3.5 KW

Wind Electric Systems, Inc.
P.O. Box 473
Santa Clara, CA 95052
(408) 243-0241
25 KW

Wind Engineering
East Airport District
P.O. Box 5936
Lubbock, TX 79417
(806) 763-3182
25 KW

Wind Harvest Company, Inc.
306 Anza Avenue
Davis, CA 95616
(916) 753-2905
10 KW (Prototype)

Windmaster Corporation
55 Veterans Boulevard
Carlstadt, NJ 07072
(201) 933-3338
55 KW

Wind Power Systems, Inc.
8630 Production Avenue
Building A
San Diego, CA 92121
(714) 566-1806
40 KW

Windtech, Inc.
P.O. Box 837
Glastonbury, CT 06033
(203) 633-7582
20 KW, 40 KW

Wind Tech Ltd.
Highway 9, West Box 396
Armstrong, IA 50514
(712) 864-3990
10 KW, 20 KW, 40 KW, 60 KW
(All Prototypes)

Windworks, Inc.
Route 3, Box 44-A
Mukwonago, WI 53149
(414) 363-4088
9 KW

WTG Energy Systems, Inc.
251 Elm Street
Buffalo, NY 14203
(716) 856-1620
300 KW

ANEMOMETER MANUFACTURERS

Aeolian Kinetics
P.O. Box 100
Providence, RI 02901

Aircraft Components
700 North Shore Drive
Benton Harbor, MI 49022

Atmospheric Research and Technology
6040 Verner Avenue
Sacramento, CA 95841

Aviation Electric Ltd.
200 Laurentien Boulevard
P.O. Box 2140 St. Lourant
Montreal, Quebec H4L 4X8

Bendix Environmental Science Division
1400 Taylor Avenue
Department #81
Baltimore, MD 21204

California Instrument, Co.
72 Dorman Avenue
San Francisco, CA 94124

Clean Energy Products
3534 Bagley N.
Seattle, WA 98103

Climatronics Corporation
1900 Point West Way, Suite 171
Sacramento, CA 95815

Climet Instruments Co.
1320 West Colton Avenue
Redlands, CA 92373

Dale-Dahl Associates
2363 Boulevard Circle
Walnut Creek, CA 94595

Danforth
Div. of the Eastern Co.
500 Riverside Industrial Parkway
Portland, ME 04103

Davis Instrument Mfg. Co., Inc.
513 E. 36th Street
Baltimore, MD 21218

Dwyer Instruments, Inc.
Box 373
Michigan City, IN 46360

APPENDIX 2

Edmund Scientific Co.
1986 Edscorp Building
Barrington, NJ 08007

Electric Speed Indicator
12234 Triskett Road
Cleveland, OH 44111

Handar
1380 Borregas
Sunnyvale, CA 94086

Helion, Inc.
Box 445
Brownsville, CA 95919

Kahl Scientific Instrument Corp.
Box 1166
El Cajon, CA 92022

Lund Enterprises, Inc.
1180 Industrial Avenue
Escondido, CA 92025

Maximum, Inc.
42 South Avenue
Natick, MA 01760

M. C. Stewart Co.
Crosby Road
Ashburnham, MA 01430

Met 1
615 Southeast Street
Grant Pass, OR 97526

Meteorology Research, Inc.
P.O. Box 637
Altadena, CA 81001

Natural Power, Inc.
Francestown Turnpike
New Boston, NH 03070

Parkway Energy Products
22 Parkway Road
Suite 2
Brookline, MA 02146

Physical Sciences Inc.
30 Commerce Way
Woburn, MA 01801

PSI Inc.
500 Cardigan Road
P.O. Box 43394
St. Paul, MN 55164

Second Wind, Inc.
2000 Massachusetts Avenue
Cambridge, MA 02140

Sign X Laboratories, Inc.
Stetson Road
Brooklyn, CT 06234

R. A. Simerl Instrument Division
238 West Street
Annapolis, MD 21401

Solar Plexus
2330 St. Mathews Drive
Sacramento, CA 95825

Sybron-Taylor Instruments
Arden, NC 28704

Systron Donner
Weather Measure Div.
3213 Orange Grove Avenue
North Highlands, CA 95660

TALA, Inc.
Route 5, Box 25
Amherst, VA 24521

Teledyne Geotec
3401 Shiloh Road
Garland, TX 75041

Texas Electronics, Inc.
5529 Redfield Street
Box 7151 Inwood Station
Dallas, TX 75209

Thermax, Canada Limited
39 Main Street
Vanleek Hill, Ontario K0B 1R0
Canada

Trade-Wind Instruments
10823 12th Avenue, NE
Dept. BH
Seattle, WA 98125

Weather Measure Corp.
3213 Orange Grove Avenue
North Highlands, CA 95660

Weather and Wind Instrument and Equipment Company
734 E. Hyde Park Boulevard
Inglewood, CA 93002

Weathertronics, Inc.
2777 Del Monte
West Sacramento, CA 95691

Westberg Manufacturing Co.
3400 Westach Way
Sonoma, CA 95476

Robert E. White Instruments, Inc.
51 Commercial Wharf
Boston, MA 02110

Wind Engineering Consultants
3421 Adams Avenue
San Diego, CA 92116

R. M. Young Company
2801 Aero-Park Drive
Traverse City, MI 49684

GLOSSARY

Airfoil	a curved surface designed to create lift as air flows over it. The blades on lift-type wind machines are airfoils.
Alternating current (AC)	electrical current that reverses the direction of flow at periodic intervals.
Anemometer	an instrument to measure wind speed.
Appliance power	the energy requirements necessary for domestic and small industrial uses.
Asynchronous-type generator	a type of generator, including DC generators and alternators, whose speed can vary over a wide range. It is used in battery-charging systems or, using an inverter, can be connected to the utility grid. Wind machines using this type of generator are characterized by a rotor whose speed varies considerably with wind speed.
Avoided cost	the costs a utility avoids because of the power contributed by small producers and cogenerators; includes savings on fuel and postponed or cancelled construction; required by federal law to be the basis for rates that utilities pay for electricity they buy back from small producers and cogenerators.
Blade	the part of a wind machine rotor with an aerodynamic surface designed to extract energy from the wind.
British thermal unit (Btu)	the amount of heat it takes to raise the temperature of one pound of water one degree Fahrenheit (°F).
Coning	a fixed orientation in which the blades of a wind machine are inclined in slightly toward each other, such that the rotor traces a conical rather than a disc shape during operation. This can reduce the blade loads and help orient the machine into the wind.
Cut-in speed	the wind speed at which a wind machine begins to produce power.
Cut-out speed	the wind speed at which a wind machine shuts down to avoid damage from high winds.
Darrieus rotor	a vertical-axis rotor with long, thin blades shaped like loops connected at the top and bottom of the axle; looks like an eggbeater.
Direct coupling	a wind machine system where a water pump is attached to both an electric motor and a wind machine.
Direct current (DC)	electrical current that flows along a wire in one direction.
Diurnal	daily; used to describe wind speed patterns that vary regularly with the time of day.
Downwind	a wind machine whose rotor is downwind of the tower. The wind first passes the tower and then the rotor.
Electrical energy	in wind, the conversion of wind power to mechanical energy to drive a turbine to produce electricity.

GLOSSARY

EPF	"energy pattern factor," the ratio of the cube of the mean wind speed to the sum of the cubes of the discrete wind speeds. Useful as a guide to how much actual conditions vary from the mean wind speed. The higher the EPF, the greater the variance.
Feather	to rotate the blades to change their pitch angle and reduce lift, so that rotor speed is controlled.
Flapping rotor	a rotor whose blades are attached to the hub bending loads in the rotor and, in some configurations, provide a simple method for controlling rotor speeds. See Coning.
Flyball governor	a device for feathering blades, using metal weights ("flyballs") mounted on each blade. When the blades turn too fast, the flyballs swing into the plane of blade rotation, twisting the blades out of proper configuration and slowing them down.
Free-standing tower	a tower which, by design and construction, is anchored so securely at its base as to require no additional bracing.
Furling	turning the rotor out of the wind to control rotor speed or provide shut-down in high wind conditions.
Governor	a system or device to control rotor speed in high winds.
Grid connected (utility interconnect) systems	wind energy systems that produce electricity capable of being connected to utility power lines.
Guyed tower	a tower anchored by three or more guy wires.
HAWT	"horizontal-axis wind turbine," a wind machine whose rotor axis is horizontal (e.g., the typical farm windmill).
Hub	the central part of a wind machine rotor to which the blades are attached; its height above ground is frequently used as a reference point for wind speeds.
Induction generator	a synchronous-type generator characterized by a relatively high slip (speed variations can be up to ten percent) and the ability to provide a considerable amount of start-up torque (rotational force). Used on many small (less than 100 kW) wind machines. See Synchronous-type generator.
Inverted block schedules	a utility measure that charges more per kilowatt to larger consumers of electricity.
Inverter	a device that converts direct current to alternating current.
Kilowatt (kW)	a measure of power equal to 1000 watts, or 1.34 horsepower.
Kilowatt-hour	a measure of electrical energy (1000 watt hours) calculated by multiplying power (kilowatts) by time (hours). For example, ten kilowatt hours are equivalent to consuming or producing two kW for five hours, or five kW for two hours. An average American household uses 750 kWh per month.
Life-cycle cost	in wind energy, a method of comparing the total cost of owning a wind system for the life of that system with the total benefits derived from the WECS.

GLOSSARY

Lift	the aerodynamic force that operates at right angles to airflow "pulling" a rotor blade along in its path.
Load management	utility district programs designed to reduce peak load demand.
Mean power output (MPO)	average output of a wind machine at various average annual wind speeds, shown on a graph or a table. Multiplying MPO by a year's time gives annual energy output.
Mean energy wind speed	average wind speed over a given time period and at a given height.
Mechanical energy	ability of rotor motion to directly perform tasks like pumping water.
Megawatt (MW)	a measure of power equal to 1000 kilowatts or one million watts.
Nacelle	a permanent enclosure mounted on top of a wind system tower housing the generator, drive train, control system, etc.
Peak load	time of greatest demand for electricity in a utility district; determines need for new generating plants. Southern climates, for example, have peak demand during summer afternoons when air conditioners are operated at maximum levels.
Power coefficient	the ratio of power extracted by a wind machine to total power available; a measure of the efficiency of the machine.
Power conditioning	changing the characteristics of electrical power; for example, from DC to AC, or from 12 volts to 32 volts.
Power density	available wind power per unit area of the windstream, usually given in watts/square meter (W/m^2). Most good wind sites have power densities above 100 W/m^2. Power density varies directly with the cube of wind speed, meaning that small increases in wind speed produce even larger increases in available power.
Power output	the useful power produced by a wind system at a given wind speed.
Process heat	describes the production of heat energy for any purpose.
Proxy evidence	in wind energy, the presence of bent trees or other environmental indicators of a site of probable high wind activity.
Rated power	the expected power output of a wind machine at a specified wind speed (its rated speed).
Rated speed	wind speed at which a wind machine produces its power. The rated power for different wind machines cannot be directly compared unless they are rated at the same wind speed.
Rayleigh distribution	a mathematical approximation of wind speed distribution. The Rayleigh distribution can be calculated knowing only average annual wind speed for a site, and gives a reasonable estimate of actual wind speed distribution when detailed wind data is not available.
Recorder	a device to record wind speed and other pertinent data. Available in strip chart and solid state models.

GLOSSARY

Rotor	the rotating assembly of a wind machine (blades and powershaft) that converts energy in the wind into mechanical energy.
Rotor speed (or RPM)	the operating speed at which the rotor will perform its designed task.
Savonius rotor	a vertical-axis rotor that is powered by drag forces. It looks like an oil drum cut in half, with the two curved sheets forming an S when viewed from above.
Shut-down	controlled condition in which a wind machine is not operating even though there is sufficient wind; used in case of high winds or for maintenance.
Shut-down wind speed	wind speed at which the control system will shut down the wind system. See Cut-out speed.
Sine wave	an oscillating wave signal typical of utility AC current. When connecting a WECS to the utility, it is necessary to match the wave signal of the WECS to the utility's.
Site analysis	assessment of the wind energy available at a particular location to tell if it is suitable for a wind system; also called "siting."
Solid frontal area (or high solidity order)	WECS whose rotor area is comprised of solid metal or wooden blades, typical or water-pumping windmills.
Solidity	ratio of the blade surface area in a rotor to the total frontal area (or swept area) of the entire rotor.
Square wave	a signal characterized by rapid changes in voltage. Square wave signals are unacceptable to utility interconnect.
Stand-alone inverter	inverter that operates without being linked to a utility grid for frequency and/or voltage signals. See Inverter.
Start-up	the point at which the rotor begins to rotate, but not necessarily to produce power. See Cut-in.
Sucker rod	in a water pump wind machine, the driving rod that goes from the rotor to the piston in the pump.
Survival wind speed	the maximum wind speed a wind machine can survive without damage. (The WECS may not necessarily be generating.)
SWECS	"small wind energy conversion system"; a wind system whose maximum output is less than 100 kW.
Swept area	as seen from the direction of the wind, the area that the rotor will pass over during one full rotation; used in calculating a wind machine's potential power output.
Synchronous generator	a synchronous-type generator that is often used on larger wind machines. Synchronous generators are more efficient than induction generators, but are much more expensive, especially in small sizes, and cannot supply starting torque (rotational force). See Synchronous-type generator.

GLOSSARY

Synchronous inverter — a device that converts DC to AC and must have another AC source (for example, a utility grid) for voltage and frequency reference. Its AC output is synchronous (same phase and frequency) with the outside source.

Synchronous-type generator — a motor/generator that can be connected directly to the utility grid. Because synchronous-type generators are essentially constant-speed machines, wind machines using them have a nearly constant rotor speed, regardless of wind speed. See Induction generator, Synchronous generator.

Tailvane — a fin (like an airplane rudder) mounted on a horizontal-axis wind machine to keep the rotor aimed into the wind.

Tip speed — the speed of the rotor tip.

Tip-speed ratio — ratio of the tip speed to the wind speed; used as a measure of rotor performance. Tip-speed ratios range from 5 to 15 for lift-type rotors.

Tower Shadow — aerodynamic wake resulting from the flow of air around a tower.

Turbulence — rapid, random wind speed fluctuations. Wind machines should be located to avoid turbulence caused by obstacles such as trees, buildings, etc.

Upwind — describing a wind machine whose rotor is upwind of the tower, facing into the wind. The wind first passes the rotor, then the tower.

VAWT — "vertical-axis wind turbine," a wind machine whose rotor axis is vertical. Typical of the "eggbeater" wind generator.

Watt — equals one ampere flowing through a potential difference of one volt.

Watt hour — a measure of electrical energy equivalent to one watt times one hour.

Watts per square meter — ratio of power available to windmill frontal area.

WECS — "wind energy conversion system," a machine that converts energy of the wind into usable form.

Wind chargers (or high electric generators) — thin bladed, fast rotor windmills used for electrical generation.

Wind farm — a collection of several wind machines at the same location.

Wind furnace — a wind energy system that converts wind power into useful heat.

Wind prospecting — the initial search for windy sites capable of producing cost-effective wind energy.

Wind regime — wind conditions with respect to its speed. A good wind regime for a specified location means generally strong winds.

Wind rose — a graph showing the percent of time that the wind blows from each direction, for a given location. Usually depicted as a circular graph, with 16 compass points radiating from the center.

Wind shear — the effect of surface friction on wind speed, slowing the wind closer to the ground. Unevenness or obstacles on the terrain add to this effect.

GLOSSARY

Windspeed distribution — the portion of time the wind blows at various speeds at a given site, plotted on a graph. Important for evaluating a wind site since small increases in wind speed result in large increases in energy.

Windspeed profile — a graph showing how wind speed varies with height above ground or water surface. Useful for deciding how high to place a wind machine.

Yaw axis — the vertical axis about which a horizontal-axis wind machine rotates to aim into the wind. "Yaw control" refers to control of movement about this axis.

BIBLIOGRAPHY

California Office of Appropriate Technology. *Model Ordinance for Small Wind Energy Conversion Systems.* Sacramento: OAT/California Energy Commission, 1982. 35 pp. free.

Center for Renewable Resources. *Promoting Small Power Production: Implementing Section 210 of PURPA.* Washington, DC: CRR, 1981. 49 pp. $3.00.

Eldridge, Frank R. *Wind Machines.* Second edition. The MITRE Energy Resources and Environment Series. New York: Van Nostrand Reinhold, 1980. 214 pp. $17.95.

Hirshberg, Gary. *The New Alchemy Water Pumping Handbook.* Andover, MA: Brick House Publishing Co., Inc., 1982. 141 pp. $8.95.

Hunt, V. Daniel. *Windpower: A Handbook on Wind Energy Conversion Systems,* Van Nostrand Reinhold, 1981. 610 pp. $39.50.

Koeppl, Gerald W. *Putnam's Power From the Wind.* Second edition. New York: Van Nostrand Reinhold, 1982. 224 pp. $27.50.

Marier, Donald. *Wind Power for the Homeowner.* Emmaus, PA: Rodale Press, 1981. 381 pp. $12.95, paperback.

McGuigan, Dermot. *Harnessing the Wind for Home Energy.* Charlotte, VT: Garden Way Publishing, 1978. 135 pp. $4.95, paperback.

Park, Jack. *The Wind Power Book.* Palo Alto: Chesire Books, 1981. 300 pp. $14.95.

Wegley, Harry L., Montie Orgill and Ron L. Drake. *A Siting Handbook for Small Wind Energy Conversion Systems.* Richland, WA: PNL: available from NTIS, 1978. 113 pp. $6.50. Order #PNL-2521.

Weis, Patricia and Steven Mooney, eds. *Wind Energy Information Directory.* Golden, CO: Solar Energy Research Institute, May 1980. 28 pp. free.

SOURCES

American Wind Energy Association
2010 Massachusetts Avenue NW, 4th floor
Washington, DC 20036

Pacific Northwest Laboratories
Wind Characteristics Program
P.O. Box 999
Richland, Washington 99352

National Technical Information Service
5285 Port Royal Road
Springfield, VA 22161

Solar Energy Research Institute
1617 Cole Boulevard
Golden, CO 80401

PERIODICALS

Alternative Sources of Energy
107 S. Central Avenue
Milaca, MN 56353
$15.00/year, bi-monthly

Renewable Energy News
P.O. Box 32226
Washington, DC 20007
$28.00/year, monthly

Wind Energy Digest
P.O. Box 306
Bascam, OH 44809

Wind Industry News Digest
Alternative Sources of Energy, Inc.
107 S. Central Avenue
Milaca, MN 56353
$36.00/ year, monthly

Windletter
American Wind Energy Association
1050 17th Street, N.W., Suite 1100
Washington, DC 20036
$35.00/year with membership

Photographs contributed by Joe Carter and Mush Emmons.

Typical California water-pumping application. Aeromotor. Photo by Carter. Page 8.

A 6-kWh, DC-producing wind machine. Owner-built. Photo by Carter. Page 12.

The view from atop the 60-foot tower of an 18-kWh, downwind, AC-producing wind machine. Wind Power Systems. Photo by Emmons. Page 22.

A 2-kWh, DC-producing wind machine. Dunlite. Photo by Carter. Page 36.

A 2-kWh, DC-producing wind machine. Dunlite. Photo by Carter. Page 48.

Installing a 6-kWh, DC-producing wind machine. Elektro. Photo by Carter. Page 60.